Roger Greeno

B.A. (Hons.), M.C.I.O.B., F.I.O.P.

Building construction
Level 3

Longman   London and New York

Longman Group Limited
Longman House, Burnt Mill, Harlow
Essex CM20 2JE, England
Associated companies throughout the world

Published in the United States of America
by Longman Inc., New York

First published 1985

British Library Cataloguing in Publication Data
Greeno, R.
   Building construction, level 3. —
   (Longman technician series.   Construction
   and civil engineering)
   1. Building
   I. Title
   690   TH145

ISBN 0-582-41321-4

Printed in Singapore by
The Print House (Pte) Ltd

Longman Technician Series

Construction and Civil Engineering

General Editor – Construction and Civil Engineering

C. R. Bassett, B.Sc., F.C.I.O.B.

Formerly Principal Lecturer in the Department of Building and Surveying, Guildford County College of Technology

Books published in this sector of the series:

Economics for the construction industry   *R. C. Shutt*
Design procedures Level 4   *J. M. Zunde*
Design technology Level 5   *J. M. Zunde*
Architectural design procedures   *C. M. H. Barritt*
Construction technology for civil engineering technicians   *P. L. Monckton*
Environmental science   *B. J. Smith, G. M. Phillips and M. Sweeney*
Concrete technology Level 4   *J. G. Gunning*
Construction drawing Level 1   *R. Boxall*
Design of structural elements Volume 1   *A. G. Smyrell*
Basic accounting for builders   *D. Hughes*
Structural analysis Level 5   *S. R. Mangalgiri*
Advanced building construction Volume 1   *C. M. H. Barritt*
Construction processes Level 1   *R. Greeno*
Building organisations and procedures   *G. Forster*
Construction site studies production, administration and personnel   *G. Forster*
Construction surveying   *G. A. Scott*
Construction technology Volumes, 1, 2, 3 and 4   *R. Chudley*
Maintenance and adaptation of buildings   *R. Chudley*
Building services and equipment Volumes, 1, 2 and 3   *F. Hall*
Measurement Level 2   *M. Gardner*
Measurement Level 3   *M. Gardner*
Structural analysis Level 4   *G. B. Vine*
Site surveying and levelling Level 2   *H. Rawlinson*

Contents

Acknowledgments

We are indebted to the following for permission to reproduce copyright material:

Bison Concrete Ltd for references and figures from trade publicity literature; British Standards Institution, 2 Park Street, London W1A 2BS for extracts from *British Standards & Codes of Practice*, from whom complete copies can be obtained; GKN Mills Building Services Ltd for extracts from trade publicity literature; The Controller of Her Majesty's Stationery Office for extracts from *Building Regulations 1976* and amendments.

Author's acknowledgements

I would like to express my thanks to Colin Bassett the Series Editor for his encouragement and help which enabled me to prepare this book. I am also very grateful to the publisher for providing the opportunity.

Additionally, I must acknowledge Margaret William's assistance in preparing and typing the manuscript – a task made difficult by the quality of my handwriting!

Chapter 1

Performance rating of construction techniques used in low-rise dwellings

The expected performance rating of an element of structure is frequently specified by the client or designer as an alternative to a detailed form of construction. This has the advantage of allowing choice of construction methods and encourages industrialists to develop new concepts without being tied to traditional specifications. British Standard 4949 : 1973 provides a glossary of terms based on building performance, and assessment is defined as, 'A judgement of predicted performance in use based on the comparison of test data with the performance requirement.'

Test data for many types of construction or components could usefully compare with the performance requirements of the Building Regulations, but it is important to be aware that performance under test may not provide the same results as performance in use. Many new products are used without the benefit of long-term environmental assessment. A reliable guide to their performance may be provided by the Agrément Board. This is an independent organisation, functioning with the Building Research Establishment, to investigate and examine new building products. A certificate issued by the Agrément Board indicates that a product has been satisfactorily tested for the purpose it is intended, and it provides an opinion as to whether it satisfies current legislation.

Floors

There are no special requirements for insulation or fire resistance of

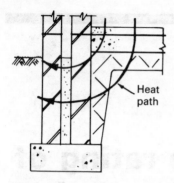

Fig. 1.1 Heat loss through solid ground floor

floors next to the ground, but where the floor is exposed to the external air, the second amendment (1981) to the Building Regulations requires the thermal transmittance coefficient to be no more than 0.6 W/m² °C. With solid floors the greatest heat loss occurs around the perimeter, as this area offers the least resistance to heat loss, (see Fig. 1.1).

Suspended floors with underfloor ventilation have a fairly constant air temperature below their surface and heat loss may be assumed the same overall. Determination of heat loss through floors is difficult due to many variable factors caused by floor area, shape and the amount and number of edges exposed. It can be expected that detached dwellings having all edges exposed will have greater heat loss through the floor than semi-detached or terraced dwellings. Examples of both suspended and solid floors are illustrated in Fig. 1.2 with typical 'U' values.

Ground floors to timber-framed housing

Timber-framed houses may be constructed with either solid concrete or suspended timber floors. The use of suspended timber complements the timber external frame and provides a superior standard of flooring than achieved with concrete. Insulation may be incorporated in the floor during construction as shown in Fig. 1.2. Although there are no special requirements for fire resistance of ground floors, where a timber-framed party wall is constructed from a suspended timber floor a fire stop must be provided below floor level. A possible construction here is shown in Fig. 1.3.

Upper floors to timber-framed housing

Upper floors are in principle the same as traditional construction except that manufacturers prefer to use chipboard or plywood sheet. Plywood is specified if the design requires stiffened or stressed skin panels (see Fig. 1.15). Sheet materials are more convenient during assembly and cheaper to purchase per unit area, but softwood

3

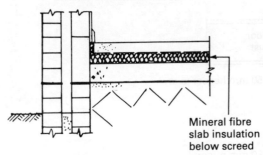

Solid floor
'U' value uninsulated, 0.9
with 50 mm insulation, 0.4

Mineral fibre
slab insulation
below screed

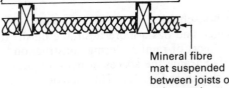

Suspended floor
U value, uninsulated, 0.9
with 100 mm insulation, 0.3

Mineral fibre
mat suspended
between joists on
polypropylene or
steel netting

Fig. 1.2 Thermal insulation of ground floors

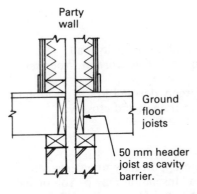

Party
wall

Ground
floor
joists

50 mm header
joist as cavity
barrier.

Fig. 1.3 Cavity barrier between suspended timber floors with timber
framed party wall

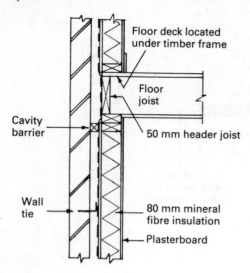

Fig. 1.4 Section through upper floor junction with timber-framed external wall

boarding is preferable to the householder, as it provides easier means of access for maintenance and has a more desirable finish. It is also more resistant to the effects of rainfall during construction. Where joists bear on the timber frame the spaces in between are closed with a header joist as shown in Fig. 1.4. This functions as a fire barrier between floor and cavity, and provides supplementary support to the upper timber frame. At the floor perimeter it is essential to provide a secure fixing for the floor deck as this performs as a horizontal diaphragm, distributing the floor loading to the external walls.

Walls

The performance requirements for traditional cavity wall construction are considered in Chapter 7, *Construction Processes, Level 1* by the same author.

The content of this section is largely devoted to the performance of external and party walls of timber-framed and cross-wall constructed dwellings

Thermal insulation

Timber-framed building provides the opportunity to achieve very high standards of thermal insulation. Insulation may be roll or batt type glass wool installed between the stud framework of the external wall. An effective vapour barrier is installed between the insulation

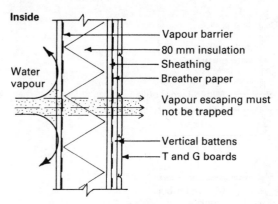

Inside

- Vapour barrier
- 80 mm insulation
- Sheathing
- Breather paper

Water vapour

Vapour escaping must not be trapped

- Vertical battens
- T and G boards

Fig. 1.5 Effect of vapour barrier and breather paper attached to timber-framed exterior wall

and the interior wall lining to minimise the passage of water vapour from within the house to the structure. The effect is shown in Fig. 1.5, and any moisture which does penetrate through joints or gaps provided for services must be allowed to escape. Also, the waterproof membrane lining the timber-frame cladding must be vapour permeable, but weatherproof to rainfall during construction and to cavity dampness in use.

The thermal insulation performance of a timber-framed external wall is well within the Building Regulation requirements of 0.6 W/m² °C. The two constructional forms shown in Fig. 1.4 and 1.5 provide 'U' values of 0.38 for brick facing and 0.41 for timber cladding. Thermal insulation between dwellings is also easily achieved by the party-wall construction shown in Fig. 1.6.

Sound insulation

Sound insulation is necessary on some sites to reduce the effect of traffic or aircraft noise. The avoidance of noise transmittance within a dwelling may also be an important design criterion, but as far as the Building Regulations are concerned the functional performance of party walls and their resistance to transmittance of airborne sound is all that needs consideration. The performance of a party wall is based on the traditionally accepted standard of a 215 mm brick wall plastered both sides. Mass is certainly an effective means of sound reduction, but not the only means and not necessarily the most effective. It is now widely accepted that a relatively light form of construction can have equally or more effective sound-absorbing properties. The Building Regulations provide a table listing the sound reduction in decibels required at frequencies ranging between 100 and 3150 Hz. This is shown graphically in Fig. 1.7 compared with the effect of twin leaf timber-framed separating walls under the same test conditions as a 215 mm thick brick wall.

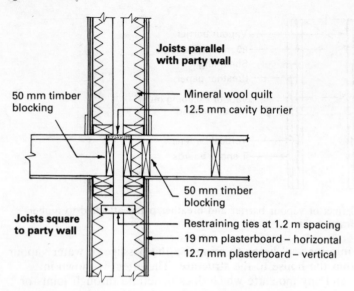

Fig. 1.6 Party-wall construction – timber-framed housing

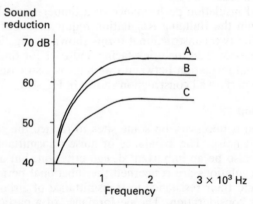

A Timber-framed party wall with 38 mm plasterboard, mineral wool insulation and 225 mm separating plasterboard.
B As A, but 32 mm plasterboard and 300 mm separation.
C Building Regulation (table to G2) requirements.

Fig. 1.7 Performance comparison of timber-framed party walls with Building Regulation requirements

Fire

External walls to dwellings must provide adequate resistance to the effects of fire to ensure the integrity of the structure in a fire situation. Part E of the Building Regulations defines the requirements, and external walls in dwellings up to three storeys must achieve at least half an hour fire resistance. This does not preclude the use of a timber frame as the basis of an external structural wall. There is however, a restriction on the use of timber as an external cladding material: it may only be used where the wall is over 1 m from the boundary. Walls of this type need only provide sufficient fire resistance from the inside of the building.

A timber-framed external wall derives its fire resistance from:

(a) the lining material,
(b) the stud framework,
(c) the insulation,
(d) the external sheathing,
(e) the cladding.

The lining material most suited to satisfying the Building Regulations is plasterboard of 12.7 mm thickness. The gypsum core is incombustible and care must be taken to ensure that joints are fully sealed and tape jointed to provide continuity of protection.

The timber stud framework is slow to deteriorate under fire, as the charred surface of timber provides sufficient protection to the core to permit loadbearing for a predictable period. Many materials could be used for the insulation between the studwork. In the interests of fire restriction it must be incombustible. Glass fibre, slagwool and rock fibre matting are ideal, but combustible insulation of expanded polystyrene, cellulose fibre and foams should be avoided.

The external sheathing is 9 mm plywood or particle board. Neither material performs particularly well under fire, but the sheathing is adequately protected by the exterior cladding (particularly if brickwork) and the internal lining. The standard timber-frame inner leaf shown in Fig. 1.4 and 1.5 more than satisfies the Building Regulation requirements for half an hour resistance. If 1 hour fire resistance is required this is simply provided by two layers of 12.7 mm plasterboard lining.

Surface spread of flame

This is another important fire criterion defined in part E to the Building Regulations. It applies to the internal surface of walls and ceilings, but fittings such as mantelpiece and windows are excluded. British Standard 476 : Part 7 : 1971 classifies materials in descending order, 0, 1, 2, 3 and 4. Untreated timber, plywood, chipboard and hardboard with a density of at least 400 kg/m^3 are rated class 3, but treatment with fire-retardant paint can improve them to class 1 or possibly class 0. The Building Regulations require ceilings in dwellings up to two storeys to have class 3 spread of flame and walls to have class 1. Houses over two storeys and flats must have ceilings and walls of class 0. These requirements are fully satisfied by employing plasterboard lining to walls and ceilings in timber-framed housing. The incombustible nature of the gypsum core provides a class 0 classification.

Building Regulation E15 (3) provides some relaxation to the routine of plasterboard by allowing a class 3 finish to walls, provided the area does not exceed the lesser of half the floor area or 20 m². So far as fire resistance of party walls are concerned, the Building Regulations are satisfied if the total resistance is 1 hour. By satisfying the resistance to airborne sound the overall thickness of plasterboard on one side is more than adequate.

Cavity barriers and fire stops

The external sheathing to the timber-framed inner leaf of dwellings is unlikely to be incombustible. Therefore, plywood or fibreboard having less than class 0 spread of flame requires the installation of cavity barriers in accordance with Building Regulation E14. Before considering their location it is useful to distinguish between cavity barriers and fire stops.

The Building Regulations specify a cavity barrier as, '. . . construction provided to close a cavity against penetration of smoke or flame . . .'. A fire stop, 'means a seal of non-combustible material provided to close an imperfection of fit between elements, components or construction in a building . . .'.

Cavity barriers in timber-framed housing may be constructed of materials specified in Building Regulation E14 section 8(b). These include:

(a) asbestos building board, minimum 9 mm thick,
(b) plasterboard, minimum 12.5 mm thick,
(c) steel of 3 mm minimum thickness,
(d) timber, minimum 38 mm thick,
(e) mineral wool blanket of 50 mm minimum thickness with wire reinforcement,
(f) cement mortar or plaster, minimum 25 mm thick,
(g) any other construction having a fire resistance of at least half an hour.

The position of cavity barriers is shown in Fig. 1.8 and Fig. 1.9 illustrates their use adjacent to a fire-resisting partition or door.

Fire stops are used mainly to seal the separating or party wall where it forms a junction with the external cavity wall and the roof covering. A fire stop is also used to seal openings for service pipes and cables where they penetrate the structure. An example of the

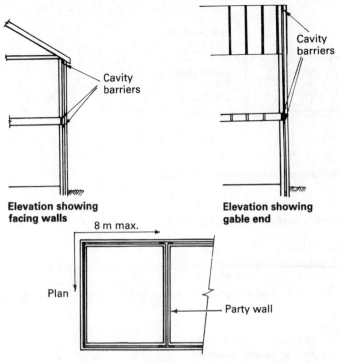

Fig. 1.8 Location of cavity barriers in timber-framed housing

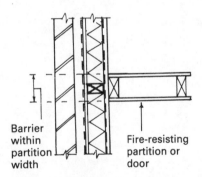

Fig. 1.9 Timber-framed inner leaf with cavity barrier adjacent to fire-resisting partition

possible treatment to a service pipe passing through a floor is shown in Fig. 1.10, and materials used in this situation could include:

(a) mineral insulation or fibre board,
(b) mineral wool reinforced with wire binding,
(c) asbestos rope,

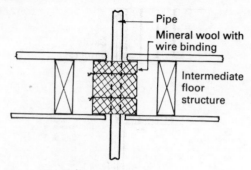

Fig. 1.10 Fire-stopping access for service pipe

(d) intumescent mastic,
(e) cement or plaster-based materials.

Separating walls

Building Regulation E8(1) requires a party wall to '. . . form a complete vertical separation between any buildings separated (including any roof spaces therein)', and regulation E5 (2) (a) states,

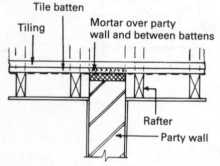

Section, where party wall meets roof

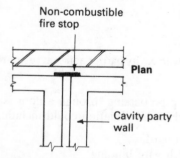

Fig. 1.11 Fire stopping to party walls

'any separating wall shall not have a fire resistance of less than one hour.'

Regulation E8 continues to specify the form of construction acceptable where the separating wall forms a junction with the roof:

E8(3): 'any separating wall which forms a junction with a roof shall be carried over the upper surface of the covering of that roof to a distance of not less than 375 mm . . .'.

This form of construction is not very attractive and alternatives are possible under regulation E8(4). For flat and pitched roofs, any part of the roof within 1.5 m of the separating wall must have a performance designation of P60*, and the covering material of the roof must be non-combustible with fire-stopping provided at the junction of wall and roof. This is shown in Fig. 1.11 for a pitched roof with tiling battens bedded solidly in mortar and the space between filled as a fire stop to the underside of the roof covering. (*See page 14).

Cross-wall construction

This form of construction has loadbearing walls arranged at right angles to the main axis of the building, hence its suitability to terraced structures. It is favoured where greater freedom of choice of external walling and finish is required. Considerable savings in construction time and materials are claimed, due to the simplicity and repetitive nature of the work. Furthermore, the infill panels forming the external front and rear cladding are of simple construction, usually timber framed and suited to factory prefabrication. The principle of cross-wall construction is shown in Fig. 1.12.

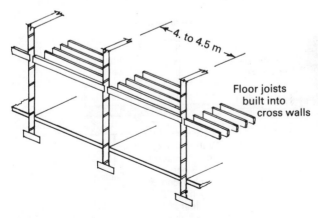

4. to 4.5 m

Floor joists built into cross walls

Fig. 1.12 Principle of cross-wall construction

Infill units must satisfy the performance requirements expected of alternative external wall forms but they are not usually expected to carry floor or roof loads. These both transfer to the cross walls, with the use of box beams in the roof space as shown in Fig. 1.14. Infill walls are expected to perform as structural bracing, to both adjacent walls and should provide sufficient rigidity and resistance to lateral loading. The floor structure will also contribute to the stability of cross walls and joists spanning between opposing walls should be built in at least 75 mm.

Continuing the cross wall to the face of the structure exposes it to penetration of dampness, in addition to problems of heat loss and condensation on the inner face near the exposed portion. Where the infill panel extends to cover to cross wall the problem is reduced, but in the absence of this technique more positive waterproofing and insulating must be incorporated. Examples are shown in Fig. 1.13.

Where infill panels are used, differential movement and shrinkage may cause the problem of defects between external and internal finishes. This is most likely where the spacing of cross walls is large and long floor spans are required. To counter these effects, stiffened or stressed skin panels should be used in roof and floor systems. Boxed plywood web beams or trussed purlins extend between rafters and ceiling joists as shown in Fig. 1.14, but for large distances support to the ceiling joists will require the additional use of hangers. In addition to providing lateral stability, these transfer most of the roof load on to the structural cross walls.

Stressed skin floor panels have a plywood skin glued to both sides of the floor joists. The plywood then performs the normal function of surface covering and additionally acts with the framing members to provide flexural strength of the whole unit. An

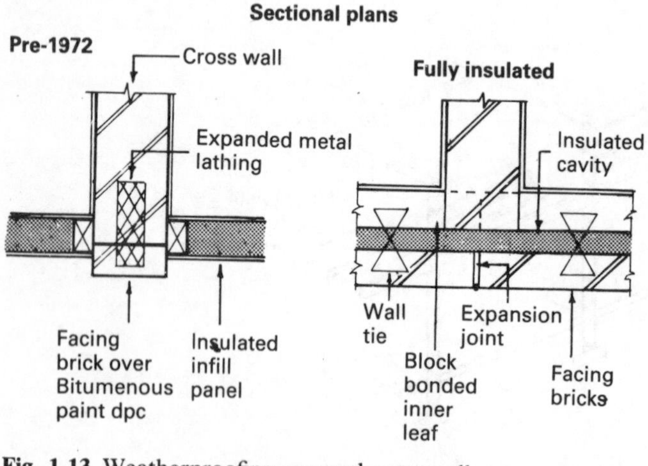

Sectional plans

Fig. 1.13 Weatherproofing exposed cross walls

alternative is stiffened panels designed to perform with the timber joists so that the plywood panel becomes an integral part of the joist to form a series of 'T' beams. Examples of both panel systems are shown in Fig. 1.15.

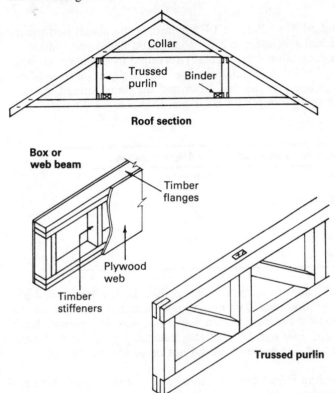

Fig. 1.14 Roof structure support with cross-wall construction

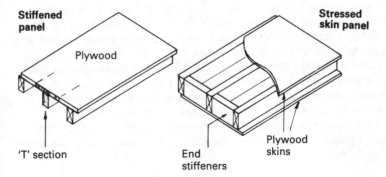

Fig. 1.15 Stiffened and stressed floor systems

Roofs

Roofs should provide a standard of performance to satisfy fire, structural, thermal insulation and durability requirements.

Fire

British Standard 476 : Part 3 : 1975 measures the actual performance data of material specimens subject to a naked flame and radiant heat. This information was originally recorded in alphabetical form, and the 1976 Building Regulations uses this designation to indicate the fire penetration by the first letter and the flame spread by the second. The relationship with BS 476 is shown in Table 1.1.

Table 1.1

Building Regulation designation	BS 476 designation
AA AB AC	P60
BA BB BC	P30
AD BD CA	
CB CC CD	P15
Unclassifiable	P5

The British Standard notation indicates that the specimen satisfied the test for ignition, and flaming was not excessive. Also, resistance to the BS test for fire penetration was for at least the period shown in seconds, e.g. P60 signifies that the material passed the preliminary ignition test and fire penetration did not occur in less than 60 seconds.

The Building Regulation requirements, and the British Standard performance ratings are represented in Table 1.2

Roof materials which cannot be classed by the BS test on account of their low softening temperature must comply with the requirements for P15 materials, except the distance from the boundary is taken as 12 m or twice the height of the building, whichever is greater. Height equals distance from mean ground level to half the vertical height of the roof.

Glass and rigid plastic sheets classified as self-extinguishing may occupy part of the roof structure if that part is over 6 m from the boundary, or, if less than 6 m, the roof is that of a garage, conservatory or outbuilding of less than 40 m² floor area.

Two examples of roof covering which adequately satisfy the British Standard requirements for P60 performance rating are shown in Fig. 1.16.

Table 1.2

Minimum acceptable BS designation	Construction limitation on buildings over 1500 m³ capacity
P60	None
P30	(a) Terraced dwellings of not more than two dwellings.
	(b) No closer than 6 m to any point on the boundary
P15	(a) Terraced dwelling of not more than two dwellings.
	(b) Not closer than 12 m to the boundary, unless the roof area is less than 3 m² and is separated from any other such part by a portion of incombustible roof covering at least 1.5 m wide, then 6 m minimum applies.
P5	(a) Terraced dwellings of not more than two dwellings.
	(b) Not closer than 22 m to the boundary.
	(c) Not over 3 m² area.
	(d) Separated from another part of the same designation roof by an area of incombustible material at least 1.5 m wide.

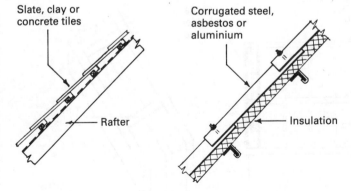

Fig. 1.16 Roof covering to P60 performance specification, re. BS 476

Structural performance

The structural requirements expected of traditional roof components is determined by Building Regulation D12(b). This requires conformity with BS 5268: Part 2: 1984 and members to be sized in accordance with schedule 6 to the Building Regulations. BS 5268 is concerned with the structural use of timber, and refers to such factors as stress grading and methods of jointing. For the determination of structural loading CP 3 Ch. V, Pt 1 : 1967 provides assessment of the expected dead and imposed loading from materials, snow and maintenance loads. With the use of CP 3 Ch. V, Pt 2 : 1972, wind loading is added to provide the total structural load. Use of schedule 6 is then possible to determine the stress grade of timber, the spacing and the size of section for the required span.

Support against wind loading to high masonry gable ends is provided by locating rafters each side of the wall and extending struts through the gable to form a ladder frame, as shown in Fig. 1.17. The loading is then restrained and distributed by tiling battens and purlin. In the absence of a purlin or where a more uniform resistance to wind is required, galvanised straps may be built into the gable and secured under two rafters as shown in Fig. 1.18. The gable complements the role of the rafters and functions as a rigid fixing to rafters at each end of the roof structure.

Trussed rafter construction is deemed to have satisfactory performance if it conforms to BS CP 112: Pt 3 : 1973, *Trussed Rafters for Roofs of Dwellings*. Compliance with the manufacturing part of the code is normally the responsibility of the truss designer and producer, but the installation aspects are the responsibility of the builder. Location to the supporting wall should be on wall plates

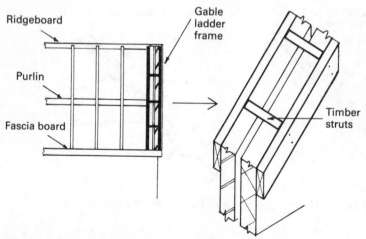

Fig. 1.17 Support between gable and roof structure

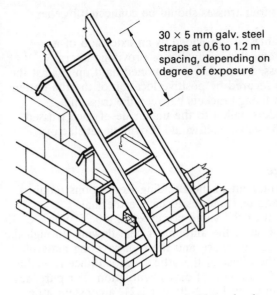

30 × 5 mm galv. steel straps at 0.6 to 1.2 m spacing, depending on degree of exposure

Fig. 1.18 Strap support between gable and rafters

at least 75 mm wide, with two 100 mm, 7 gauge galvanised wire round nails skew nailed to each end. Temporary bracing will be necessary during construction and permanent wind bracing will be needed where gable walls are of insufficient strength. The lightweight spandrel panel frequently used in timber-framed housing offers insufficient roof support, and ceiling binders in addition to the diagonal bracing shown in Fig. 1.19 will be necessary to preserve

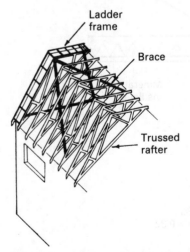

Ladder frame

Brace

Trussed rafter

Fig. 1.19 Diagonal wind bracing to trussed rafter roofs

truss stability. At least four trusses should be connected by the diagonal braces.

Where masonry gables are provided as an extension of a traditional cavity wall and this is capable of providing lateral resistance to wind, trussed rafters are positioned hard up against the inside of this wall and secured by positive location of the tile battens to the gable wall, or by fixing brackets between the truss and wall. Longitudinal binders nailed to the underside of each trussed rafter and built into the gable wall would provide a satisfactory alternative.

Thermal performance

The 1981, second amendment to the Building Regulations requires roofs of houses, flats and maisonettes to have a thermal transmittance coefficient or 'U' value not exceeding 0.35 W/m² °C. In a pitched roof this represents the amount of heat flowing through the ceiling structure, the attic air space and the roof-covering material. Precise calculation of the effect of the attic space is difficult, as this varies with pitch and the amount of eaves ventilation. For purposes of calculating the examples in Fig. 1.20 and 1.21, a constant attic space resistance of 0.16 m² °C/W is assumed. Site exposure will also have an effect and external surface resistances are taken at the norm of 0.045 m² °C/W.

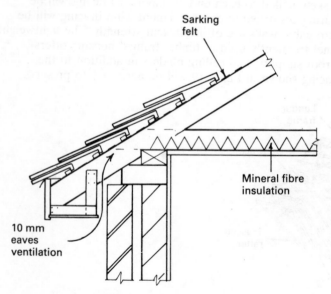

Sarking felt

Mineral fibre insulation

10 mm eaves ventilation

'U' value as shown without insulation = 1.60
'U' value as shown with 100 mm insulation = 0.32

Fig. 1.20 Traditional plain-tiled roof

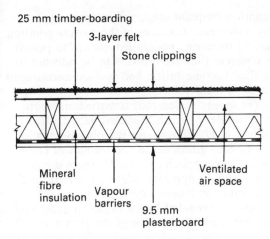

25 mm timber-boarding

3-layer felt

Stone clippings

Mineral fibre insulation

Vapour barriers

9.5 mm plasterboard

Ventilated air space

Fig. 1.21 Traditional flat-roof construction

The roof form shown in Fig. 1.20 is typical of many existing roofs and the basic 'U' value without insulation is compared with the effect of providing glass or rock fibre matting laid between ceiling joists. A similar example for a flat roof is shown in Fig. 1.21, with insulating fibre between the joists supported above the plasterboard ceiling.

Durability

The only deteriorating factors likely to affect timber in roof structures are woodworm attack or rot due to dampness. Woodworm infestation is possible in any part of the country, but the Building Regulations B3 (1) and (2), in response to known serious attacks by the house longhorn beetle, specify softwood timber treatment to all roof components used in particular areas of south-east England. Building Regulation B4 directs the nature of treatment and this may be by diffusion of sodium borate following felling and milling to produce a boric acid content after the timber is kiln dried. An alternative process for seasoned timber is immersion in a persistent organochlorine contact insecticide for not less than 10 minutes. Timber treated in accordance with BS 4072 : 1974 is also acceptable. This is based on a water-borne composition of copper, dichromate and arsenic applied by vacuum/pressure impregnation.

Dampness in roofs is possible from faults in the tiling or from condensation. The effect from either could be fungal decay of the structural timber. Roof-covering materials have a limited life and deterioration of clay tiles, and to a lesser extent slates, by delamination is very common on old roofs. As the material ages it becomes porous and, during the winter, freezing causes areas to spall and break off the tile. The decay of old nail and timber peg fixings on clay roofs is another frequent source of leakage, and structural settlement may also create fractures in the mortar pointing to ridge and verge, sufficient to cause enough disturbance to permit rain penetration. Dense concrete tiles are unlikely to be affected by spalling, but as they age they become brittle and porous, encouraged in some areas by the action of moss roots.

Condensation will occur in any sealed roof construction where provision for ventilation is omitted. British Standard 5250 : 1975, *The Control of Condensation in Dwellings*, and Building Regulation F5 (second amendment) provides guidance on the amount of eaves ventilation required. Where the roof pitch exceeds 15° there should be a 10 mm continuous opening on opposite sides of the roof, and for pitches under 15 ° a 25 mm continuous opening is necessary. Cross ventilation should also be provided by air bricks in gable ends, occupying an area of not less than 0.3 per cent of the plan roof area. Ventilation at the eaves is easily achieved by leaving the soffit board clear of the external wall as shown in Fig. 1.20.

Chapter 2

Performance and construction techniques used in low-rise commercial buildings

The style and construction of low-rise commercial buildings depends considerably on the chosen type of roof form. The structure may have conventional cavity walls, or a frame of reinforced concrete or structural steelwork. The structural performance of each is considered in Chapter 3. This chapter considers the following two-dimensional roof support frames:
(a) Structural timber
      Box beam
      Laminated timber portal frame
      Truss
(b) Reinforced-concrete portal frame
(c) Structural steelwork
      Lattice truss
      Portal frame

Structural timber

Box beams

These are illustrated in Fig. 1.14 employed as trussed purlin support for pitched roofs with cross-wall construction. They are constructed of two plywood webs with softwood flanges and intermediate struts or stiffeners. They may also taper for mono-pitch construction or bow to form arched roofing. Their major advantages over other structural roof forms are a high strength-to-weight ratio, simplicity of

22

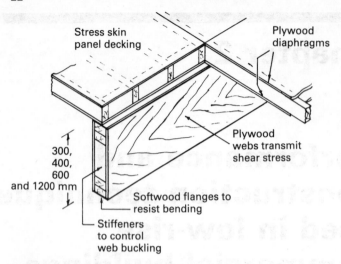

Fig. 2.1 Box beam and stress skin decking

construction, and when used with stress skin roof deck panels as
shown in Fig. 2.1 (see also Fig. 1.15) the combined strength of the
plywood webs and diaphragms offer considerable economies in solid
timber. The most economic spans are between 3.5 and 10 m.

Laminated timber

Laminated-timber structural members have a considerable history. In
modern form, timber laminates of 20 to 45 mm thickness are glued
with synthetic resins to produce an exceptionally strong unit.
Laminated-timber portal frames provide a very attractive roof
support system popular in public buildings such as community
centres, school halls, churches and swimming pools where clear
medium spans are required. They are expensive, as considerable
assembly time is required for their fabrication. However, they
provide a very attractive structure featuring varnished softwood.
Figure 2.2 illustrates the form and location of a glued, laminated
portal frame manufactured in two halves with a bolted apex
connection.

Trusses

Triangular trusses manufactured from huge timber sections
connected with iron straps, bolts and traditional mortice and tenon
joints were popular roof forms for industrial premises during the
nineteenth century. These preceded the introduction of lighter
weight steel trusses, but timber trusses were still produced and, due
to the efforts of the Timber Research and Development Association
in the early 1950s, more rationally designed timber trusses became
available for industrial and domestic use. These lattice timber frames

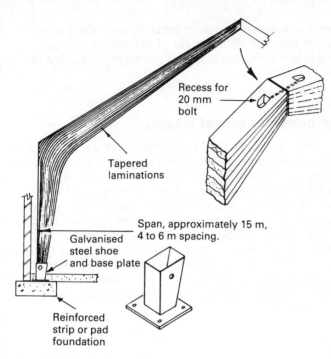

Fig. 2.2 'Glu-lam' portal frame

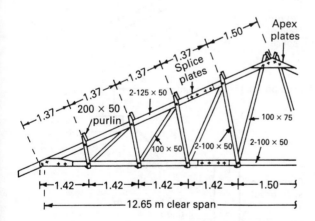

Fig. 2.3 TRADA type, industrial timber truss

are particularly useful for small to medium spans and assembly is by simply bolted connections incorporating split rings or toothed plates and plywood gussets at significant junctions. Figure 2.3 shows a typical industrial truss with timber purlins superimposed on the rafters to carry corrugated asbestos or other suitable sheet roof covering. Trusses have the disadvantage of obstructing daylight where roof lights are used, and poor fire resistance if exposed or untreated.

Reinforced-concrete portal frames

Portal frames have the advantage of providing clear unobstructed space from floor to rafter level. This convenient arrangement permits freedom for storage and production, hence the suitability and popularity of these frames for farm-building, workshop and warehouse construction. Reinforced concrete portal frames compare well, economically, with steel and are considerably cheaper than the timber equivalent. They are factory produced under quality controlled supervision in two, three or four sections, to suit spans up to 25 m. Site assembly is by simple bolted connections as shown in Fig. 2.4 with ground support provided by cast *in situ* concrete pocket pad foundations. Surface treatment is unnecessary as the unit is fire and corrosion resistant.

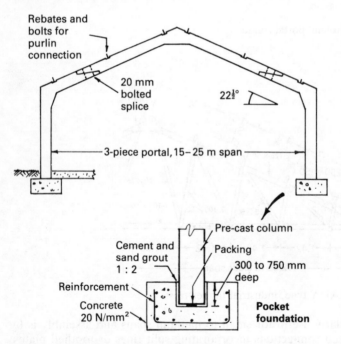

Fig. 2.4 Reinforced concrete portal frame

Structural steelwork

Lattice trusses

Since the commercial conversion of iron to steel in the latter part of the nineteenth century, steel sections have been successfully combined to produce economic, lightweight triangular roof trusses capable of spans up to about 25 m. Figure 2.5 shows various truss forms composed of standard angles for rafters, ceiling ties and internal bracing. Heavy lines represent members in compression, and thin lines those in tension.

Truss spacing using economic angle sizes is between 3 and 6 m, and roof pitch is between $22\frac{1}{2}°$ and $30°$ to suit asbestos cement or

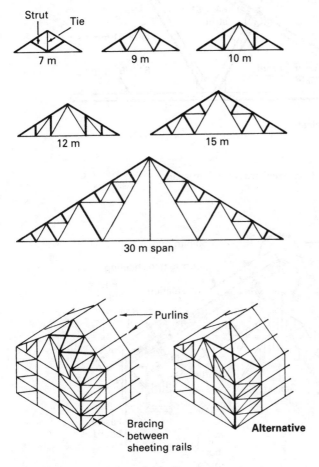

Fig. 2.5 Steel trusses and windbracing

similarly profiled plastic or metal sheeting. Trusses are factory prefabricated from standard angle sections welded or bolted to gusset plates with centre lines of components coinciding at intersections. Figure 2.6 shows the construction of a typical fink pattern truss with details of apex and eaves in Fig. 2.7.

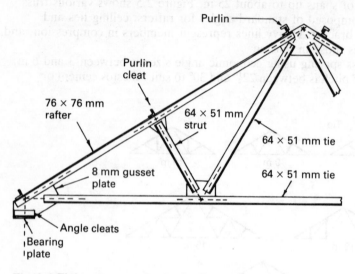

Fig. 2.6 Fink pattern steel truss for 7 m span

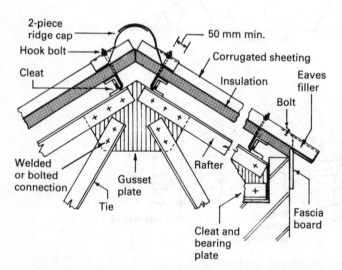

Fig. 2.7 Steel truss, ridge and eaves

Portal frames

Steel portal frames are composed of either a horizontal beam or pitched rafters joined rigidly at the eaves to a stanchion. Most are a continuous frame fabricated from universal sections with the rafter components inclined at $22\frac{1}{2}°$ for corrugated steel or asbestos roofing. Figure 2.8 shows possible construction using welded and bolted connections for frames spanning up to 25 m with spacing between 3 and 7.5 m depending on roof load. The alternative use of lattices and open web beams considerably increases the span potential as the deeper sections offers greater resistance to deflection, provided the roof load is minimised by careful choice of covering material. Figure 2.9 shows possible flat and cambered beams capable of spans up to 40 m.

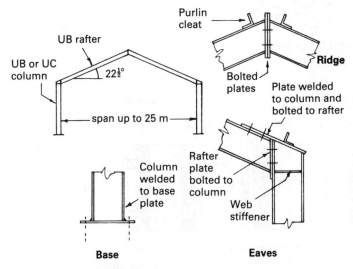

Fig. 2.8 Steel portal frame

Insulation

During 1979, the first amendment (Part FF) to the Building Regulations introduced new mandatory requirements for thermal insulation of all buildings other than private dwelling houses. This amendment improved the provisions of the Thermal Insulation (Industrial Buildings) Act and will eventually replace it. Part FF4 effectively controls the amount of window and rooflight openings and provides maximum acceptable 'U' values for walls, roofs and floors exposed to the external air. Factories, warehouses and general storage areas are required to have each element constructed with sufficient thermal insulation to provide a 'U' value of not greater

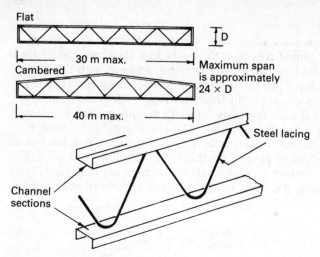

Fig. 2.9 Steel lattice joist

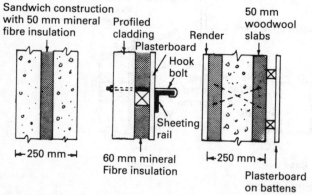

Fig. 2.10 Wall construction

than 0.7 W/m² °C, and shops, offices, institutional and public buildings constructed with elements having a 'U' value not exceeding 0.6 W/m² °C. Figures 2.10, 2.11 and 2.12 show several possible methods of achieving these requirements for walls, roofs and floors respectively.

Brick infill panel walling

Non load-bearing infill brick and block cavity walls are required to provide an attractive finish to framed structures, adequate thermal

29

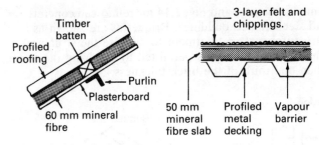

Fig. 2.11 Roof construction

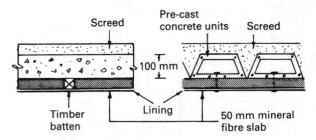

Fig. 2.12 Floor construction over the external air

insulation, resistance to rain penetration and wind pressure whilst still remaining sufficiently self supporting and stable.

The stability and strength of the frame may be enhanced by the infill wall, but for design purposes the frame is considered independently, therefore settlement of the frame and differential thermal movement must be absorbed between the two elements. Possible constructional forms are shown in Fig. 2.13 for single-storey

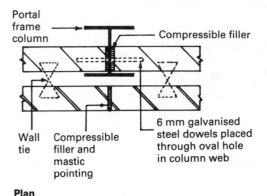

Plan

Fig. 2.13 Cavity wall construction with a steel column

steel portal-framed structures and Fig. 2.14 for reinforced concrete or concrete-clad steel-framed buildings. Briquettes are shown as a possible non-structural surface treatment to disguise the structural frame where the outer brick wall is permitted to overhang the frame by not more than one third of the brick thickness.

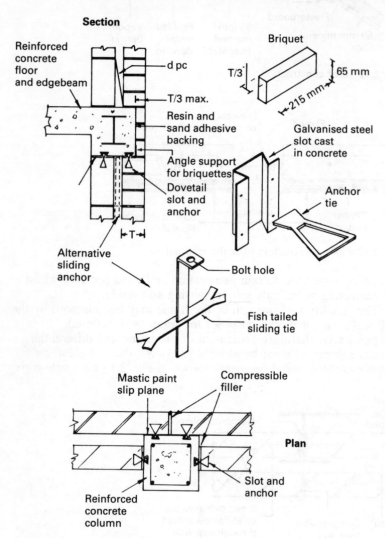

Fig. 2.14 Cavity wall infill panels to framed buildings

Chapter 3

Structural safety requirements

Scaffold

Many building site accidents occur because of the incorrect use of scaffold. This accounts for the extensive legislation and advisory documents recommending the safe use of this means of access. This section is not devoted specifically to safety, but the implications of the information and illustrations used will obviously have direct significance.

Frame construction with scaffold tubing is the most common means of providing building access. Many specialised and simplified access frames exist, but traditional scaffold systems are frequently preferred for reasons of cost and flexibility in use. The basic components are metal scaffold tube to BS 1139 : 1964, and timber boards to BS 2482 : 1981. Scaffold tube may be galvanised or black mild steel of welded, seamless or cold-rolled close-jointed manufacture with an external diameter of 48.4 mm. Lighter weight, seamless aluminium tube of the same diameter is also acceptable and both materials are available in lengths up to 6.3 m.

Scaffold boards, otherwise known as planks or battens, are manufactured from sawn softwood in widths of 225 mm. Thickness may be 38 mm, 50 mm and 63 mm. Ends of boards must be protected from splitting and may be chamfered for a distance of up to 25 mm. End protection is either by galvanised or sheradised mild steel hoops secured by nails, or with galvanised sheet steel nailplates. Both methods are illustrated in Fig. 3.1. Boards of acceptable

End hoops

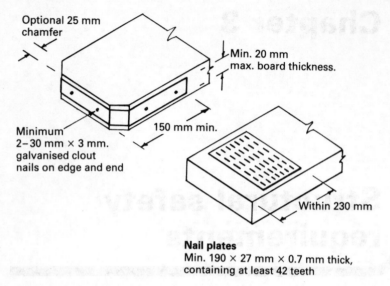

Optional 25 mm chamfer

Min. 20 mm max. board thickness.

Minimum 2–30 mm × 3 mm. galvanised clout nails on edge and end

150 mm min.

Within 230 mm

Nail plates
Min. 190 × 27 mm × 0.7 mm thick, containing at least 42 teeth

Fig. 3.1 Treatment to scaffold board ends

quality are assessed by visual or machine grading. The standards are illustrated in Fig. 3.2 and contain the following factors.

Knots
Face knots. These are measured on both sides and at right angles to the length of the board. If any one knot exceeds 75 mm or the sum of the measurement exceeds 150 mm, the board is rejected.

Edge and splay knots. Maximum three-quarters of the board thickness, in total.

Arris knots. Edge measurements, maximum three quarters of the board thickness in total and face measurement, maximum 120 mm total.

Fissures and resin pockets
Maximum length 225 mm on the face of a board where the depth exceeds one third of the board thickness. Edge fissures of depth over one third the board thickness to be no longer than twice the board thickness.

Wane
Maximum aggregate length 25 mm on either face. No wane to occur under end hoops.

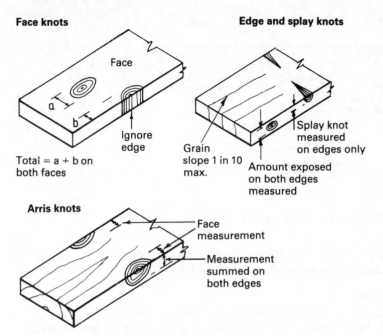

Fig. 3.2 Measurement of knots for visual grading of scaffold boards

Distortion
Bow. Maximum half the board thickness in any 3 m length.

Spring. Maximum 10 mm in any 3 m length.

Twist. Maximum 10 mm over full width of board in any 3 m length.

Cup. Maximum 5 mm.

Slope of grain
1 in 10 on face or edge.
Boards satisfying the BS requirements are marked on the end hoops
or a separate identity plate to show:

1. The BS number.
2. The supplier's name.
3. An M or V to indicate machine or visual grading.
4. The word 'support' followed by the maximum span in metres,
   (see Fig. 3.3).

Tubular members

Scaffold systems are composed of metal tubes performing different
structural functions. The main components include standard, ledger,

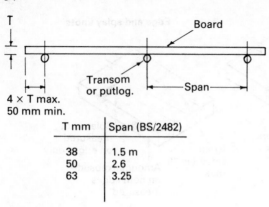

Fig. 3.3 Scaffold board support

T mm	Span (BS/2482)
38	1.5 m
50	2.6
63	3.25

brace, transom and putlog. Their location is shown in the putlog and independent scaffold illustration in Figs 3.4 and 3.5

Standards

These are the vertical components transmitting loads from the working platform to the baseplates and soleplates at ground level. Spacing varies depending on the use and purpose of the scaffold, but for normal construction and maintenance purposes will be between 1.75 m and 2.5 m. They should be vertical and joints in adjacent standards staggered and arranged to occur as close as possible to a ledger.

Ledgers

Ledgers are clamped at right angles to standards and arranged horizontally for the full length of the scaffold. They provide support to other scaffold members, in particular transoms and putlogs. Joints in adjacent ledgers should be staggered to avoid occurrence in the same bay. Sleeve couplings are preferred to joint pins and should occur within one quarter to one third of the distance between standards, never in the middle.

The height of lift or spacing will be between 1.375 m for bricklayer's use, to 3 m on multi-storey buildings; however, 1.8 m to 2.1 m is more usual. The first lift is generally about 2.6 m above ground level to provide clear space for movement of operatives and materials.

Braces

Diagonal ledger bracing should occur at right angles to the building at alternate pairs of standards for the full height of the scaffold. It will form an angle between 30° and 60° to the horizontal depending on the width of scaffold and height of lifts between ledgers.

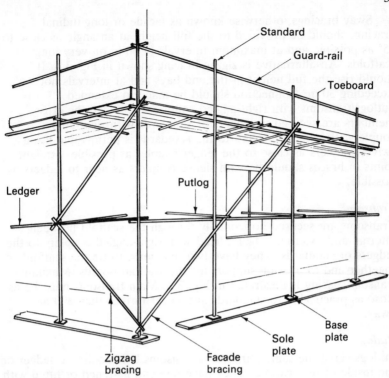

Fig. 3.4 Putlog scaffold

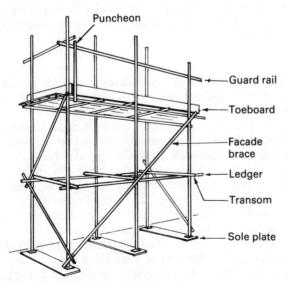

Fig. 3.5 Independent scaffold

Sway bracings, otherwise known as facade or longitudinal bracing, should be provided to the full height at an angle as close to 45° as possible and at maximum intervals of 30 m on very long scaffolds. An alternative is zigzag bracing shown in Fig. 3.4. It should rise the full height in the end bays and at intervals not exceeding 30 m. Connection should be made to extended putlogs/transoms with right-angled couplers, provided the cross members are secured to ledgers or standards with right-angled couplers. Bracing connected to the standards is permissible using swivel couplers as close to the ledger coupler as possible. Straight joints in braces should be with sleeve couplers as near to ledgers as possible.

Transom

Transoms are secured horizontally through the scaffold between internal and external ledgers, fixed with right-angled couplers to the ledgers or standards. They have two functions, to tie the scaffold together and to provide support for the scaffold boards. Maximum transom spacing is illustrated in Fig. 3.3. Main transoms must be as close as practicable to standards and never further than 300 mm away.

Putlog

Putlogs serve the same purpose as transoms but require no ledger on the inside of the scaffold. They have one end flattened or fitted with an adaptor for bearing on 75 mm of the bed joint of brickwork. Spacing and location is the same as required for transoms. They should be secured horizontally and coupled to standards or ledgers with right-angled couplers. Intermediate putlogs may be secured to ledgers with putlog couplers.

Putlog scaffold

This scaffold system is often known as a bricklayer's scaffold, as the lift heights are at chest level and only the highest lift can be used as a working platform. It is not recommended for use with tall structures and access to buildings over three storeys is more secure with an independent scaffold. Inner support is provided by the wall and if the loading is likely to approach 2.5 kN/m² per day (one operative, mortar and about 175 bricks) then the single row of standards should not exceed 1.8 m spacing along the scaffold. They should be arranged at 1.3 m from the wall, permitting five boards to be comfortably positioned. Ledgers are spaced at approximately 1.4 m vertically and remain in place as the scaffold increases in height. Putlogs may be removed as the scaffolding rises, but at least one must remain secured within 300 mm of a standard in each bay.

Facade bracing is required to the full height, inclined at between 30° and 45° to the horizontal. Zigzag bracing in the end bays and at

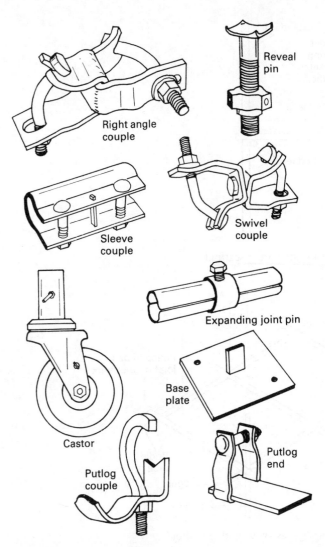

Fig. 3.6 Scaffold fittings

30 m intervals may be provided as an alternative. With a putlog scaffold, tying through the building is essential to restrain movement towards or away from the building. Ties are required in alternate lifts and at a maximum of 6 m intervals along the scaffold. All connections are with right-angled couplers and a typical detail is shown in Fig. 3.7. Openings not required for ties are by-passed with a bridle tube as detailed in Fig. 3.8.

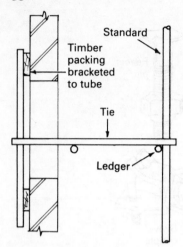

Fig. 3.7 Reveal tie for putlog scaffold

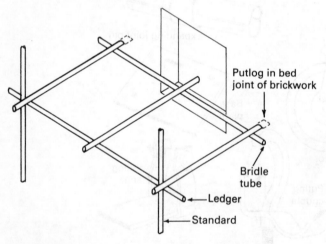

Fig. 3.8 Opening by-pass with putlog scaffold

Independent scaffold

These scaffolds contain two rows of standards connected longitudinally with ledgers and transversly with transoms. The design of scaffold and arrangement of components will depend on the anticipated loading. For maintenance and light duty work, a single platform with loading not exceeding 0.75 kN/m² per bay (one operative, tools and a small quantity of materials) require standards at 2.7 m spacing. For more general building purposes with up to four levels operating, a load of 1.75 kN/m² per bay on each platform (one operative, mortar and about 125 bricks) requires standards at 2.1 m spacing.

The inner row of standards should be as close as practicable to the building face. Where projections such as window sills prevent this, a space not exceeding 300 mm may be provided between the working platform and the wall. This permits a single board to prevent debris dropping, to be placed between the wall and the inner standards on transoms that are no further than 50 mm from the wall, as shown in Fig. 3.9.

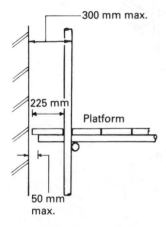

Fig. 3.9 Scaffold wall clearance

With a four-board working platform width the standards will be about 1 m apart and for a five-board platform, 1.24 m. Ledger spacing of 1.8 m to 2.1 m provides headroom at all platforms and transoms are spaced at 1.2 m to 1.5 m, depending on the quality of boards, with one in each bay no further than 300 mm from a standard, secured with a right-angled coupler. Putlog couplers may be used for securing intermediate transoms. Bracing is essential, and diagonal bracing is provided at alternate pairs of standards with either facade zigzag bracing to ensure longitudinal stability.

A height of up to about 45 m is possible with these scaffolds, provided they are securely tied at intervals not exceeding 6 m horizontally and 4 m vertically. The most satisfactory method of tying is shown in Fig. 3.10 with the through tie attached to both ledgers and restrained behind the wall. The detail in Fig. 3.11 is an alternative showing the tie tube secured to a reveal pin.

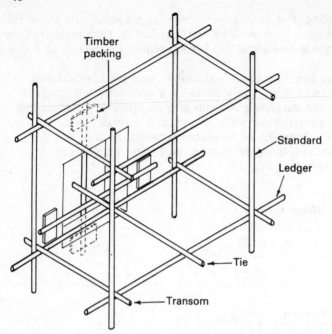

Fig. 3.10 Reveal fixing for independent scaffold

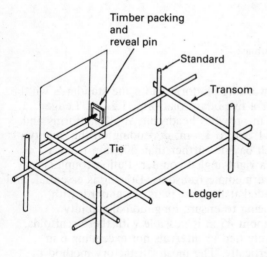

Fig. 3.11 Reveal tie fixing of independent scaffold

Working platforms

Additional safety requirements are contained in BS 5973 : 1981, *Access and Working Scaffolds*, and the 1982 Construction (Working Places) Regulations. Some of the most significant legislation affects the use of platforms, guard rails, toeboards and ladders. Platforms must be fully boarded to provide safe access. In normal circumstances it is sufficient to allow the boards to bear directly on transverse support without bracketing, but in exposed situations boards must be restrained to prevent displacement by wind forces. Clips of the type shown in Fig. 3.12 would be suitable as their projection is insufficient to cause a hazard.

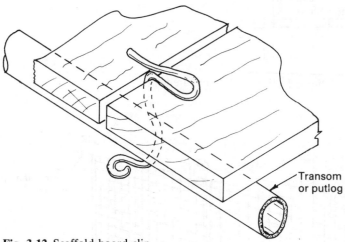

Transom
or putlog

Fig. 3.12 Scaffold board clip

Platform widths

1. General use or as a footing only, minimum 600 mm, 3 boards.
2. For deposit of material, minimum 800 mm, 4 boards, with 430 mm clear passage.
3. For use with conveyance of materials, at least 600 mm clear passage.
4. To support trestles or similar higher platform, minimum 1050 mm, 5 boards.
5. Used to work masonry, minimum 1300 mm, 6 boards.

Guard rail and toeboard

All working platforms in excess of 2 m above ground level must be provided with suitable guard rails and toeboards as shown in Fig. 3.13. Guard rails are fitted to the inside of the standards with right-angled couplers, at a height not less than 915 mm and not more

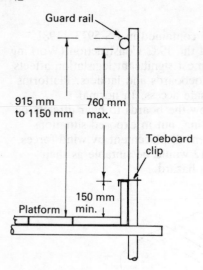

Fig. 3.13 Working platform safety requirements

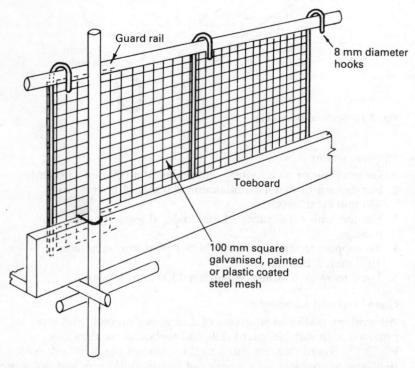

Fig. 3.14 Side protection to working platforms

than 1150 mm above the platform. The open space between the underside of the guard rail and the top of the toeboard must not be more than 760 mm. Toeboards are scaffold boards secured upright to the inside of standards with clips. The regulations require a height of at least 150 mm, so a standard scaffold board adequately fulfils this. Boards butt against each other with joints as near to standards as possible.

Side protection to platforms is recommended in BS 5973 in the form of mesh or steel fencing between toeboard and guard rail. This is to prevent people or large objects falling into the space. The fencing should be combined with the guard rail and toeboard as shown in Fig. 3.14, and have holes not exceeding 100 mm square or of 50 mm maximum width slots as the smaller dimension.

Ladders

Ladders must be securely tied to their upper resting place, but if this is impractical a secure fixing should be provided at the lower end. A firm level footing is essential and if lower or upper support cannot be provided, a person must foot the bottom rung to prevent the ladder slipping.

Safe fixing is illustrated in Fig. 3.15 with the minimum 1.05 m projection above the place of landing and an inclination of about 4 : 1.

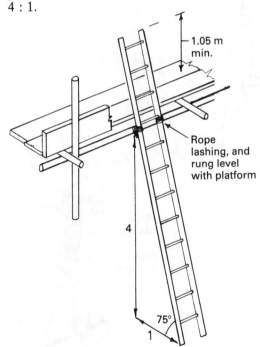

1.05 m min.

Rope lashing, and rung level with platform

4

75°

1

Fig. 3.15 Safety requirements for ladders

Gin wheels

Gin wheels are a popular simple method of hoisting fittings and materials up scaffolds. They are fitted with a hook or a ring, the latter being particularly useful on a scaffold as it slips over the projecting support tube to be located with a coupler either side. With an independent scaffold the gin wheel support tube must be secured to two standards with loadbearing right-angled couplers, and if it projects more than 300 mm it must be fitted with a diagonal brace to avoid overstressing the tube. Older gin wheels have a hook which must be 'moused' to prevent the hook jumping off the support lashing. The maximum recommended imposed load is 50 kg and the hoisting rope must be sound enough for the purpose. Suitable rope is often marked with a copper identity ring and lifting hooks are purpose made, therefore bent reinforcement bars or other makeshift devices should not be used. The application and safe use of a gin wheel is shown in Fig. 3.16.

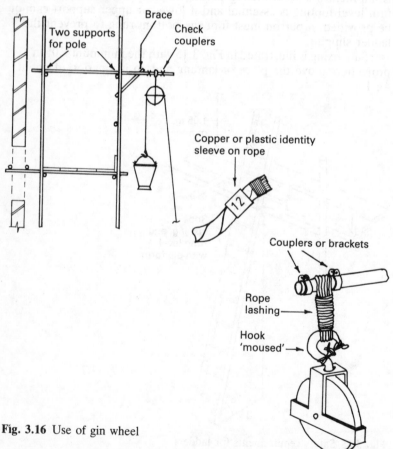

Fig. 3.16 Use of gin wheel

Mobile and free-standing scaffold

Mobile scaffolds with castor wheels attached to the base and free-standing scaffolds are either purpose made or prefabricated from standard scaffold tube and fittings. The maximum height to the platform should not be more than three times the minimum base dimension when used externally and three and a half times for internal use. Movement of mobile scaffolds is by pushing and pulling at the base as movement elsewhere could cause overturning. A firm level base is always essential. Unless the joints between scaffolding units are designed to withstand twisting forces, diagonal bracing must be provided in the horizontal and vertical plane. Fig. 3.17 shows the safety requirements necessary with these scaffold variations.

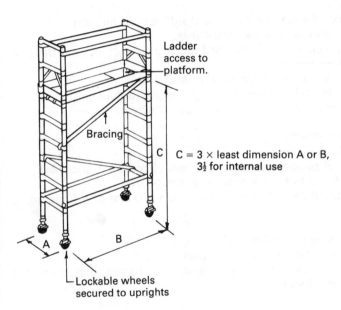

Fig. 3.17 Access tower

Scaffold check list

Ties

1. Should be adequate in number and sound in construction.
2. There should be one in every other lift.
3. One in every 6 m along the scaffold.
4. Fifty per cent of ties to be through, the remainder reveal fixing.
5. Extra ties used where fans and sheetings can respond to wind forces.

Ledgers

1. Must be truly horizontal.
2. Joints staggered and not in the same bay.
3. Sleeve couplers are preferred to joint pins.
4. Ledger bracings should be secured to alternate pairs of standards, not to hand rails, etc.

Standards

1. Must be truly vertical to avoid bowing.
2. Joints staggered so that there are never two in the same bay.
3. Baseplate should be provided under every standard.
4. Spacing should be in accordance with loading.

Soleplates

1. Should be of sound condition and sufficient in quantity.
2. Extend under at least two standards.
3. Extend a minimum of one third the bay distance beyond the last standard.
4. Joints to occur within the middle third of a bay.
5. If used internally, arranged at right angles to floor joists.
6. Are not required if base is solid, e.g. concrete.

Putlogs and transoms

1. Should be horizontal.
2. Flat end of putlogs bearing on 75 mm of bed joint brickwork.
3. Spacing is 1.5 m for BS graded 38 mm boards.
4. These should be one in each bay within 300 mm of a standard.

Boards

1. Should comply with BS 2482.
2. They should have adequate support.
3. There should be a overhang, minimum of 50 mm, maximum four times board thickness.

Platforms

1. Must be fully boarded.
2. Rubbish and surplus materials must not be allowed to accumulate.
3. Toeboards and hand rails provided throughout where platform is over 2 m above ground level.

Sway bracing

1. Should be fitted to full length of scaffold.
2. As close to 45° as possible.
3. Not more than 30 m apart.

Ladders
1. Not to be used with missing or damaged rungs or split stiles.
2. Must be adequately secured.
3. Extend at least 1.05 m above landing.
4. Landing points not more than 9 m apart vertically.
5. Landing points complete boarded out with hand rail and toeboard.
6. Used at correct angle to horizontal.

Other factors to be checked
1. Hoist towers: loadbearing couplers used throughout.
2. Improper removal of parts.
3. Scaffold board movement.
4. Effect of nearby excavations.
5. Effect of hoist or traffic vibrations
6. Damage by impact from plant.
7. Incorrect use, e.g. formwork support.
8. During assembly and dismantling a sign displayed, WARNING! DO NOT USE SCAFFOLD.

Structural stability requirements for loadbearing masonry walls

Part D8 to the Building Regulations requires masonry walls to be stable under all types of load, and schedule 7 provides guidance to the design and general construction of walls mainly for use with dwellings up to two storeys. These are reviewed in Chapter 7 of *Construction Processes, Level 1*.

This section considers in more detail the effect of BS 5628 : Part 1 : 1978, on lateral and horizontal stability requirements and slenderness factors in design of loadbearing masonry walls and columns.

Slenderness

In order to withstand the effect of dead, imposed and windloads, and for the structure to remain reasonably integral following accidental damage, a wall or column should be constructed to a suitable standard of dimensional stability. The slenderness ratio is a simple method of relating height and thickness to stability. This is expressed as the ratio of effective height or length to the effective thickness, i.e. $\frac{\text{effective height}}{\text{effective thickness}}$. The maximum slenderness ratio is 27 except for walls of less than 90 mm thickness in buildings over two storeys, where it should not exceed 20.

Effective Height

The effective height of a wall is taken as:

1. Three-quarters of the clear distance between lateral supports which provide enhanced horizontal resistance to lateral movement; or
2. The clear distance between lateral supports which provide simple horizontal resistance to lateral movement.

Examples of enhanced and simple horizontal resistance are illustrated in Figs 3.18 and 3.19.

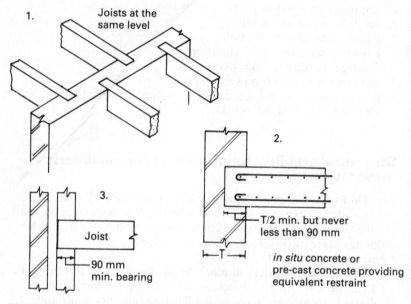

Fig. 3.18 Examples of enhanced horizontal resistance to walls

Enhanced horizontal resistance
1. Floor or roofs of any constructional form spanning onto both sides of the wall or column at the same level.
2. An *in situ* concrete floor or roof bearing at least 90 mm or no less than half the wall or column thickness.
3. A house up to three storeys with timber joists bearing at least 90 mm on one side of the wall.

Simple horizontal resistance
This applies to houses having timber joists spaced at no more than 1.2 m apart bearing on suitable joist hangers of the type shown in Fig. 3.19. It also applied to buildings having concrete floor

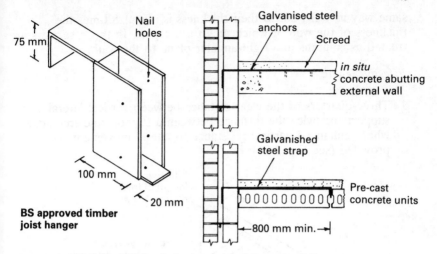

Nail holes

75 mm

100 mm

20 mm

BS approved timber joist hanger

Galvanised steel anchors

Screed

in situ concrete abutting external wall

Galvanised steel strap

Pre-cast concrete units

800 mm min.

Fig. 3.19 Examples of simple horizontal resistance to walls

construction, and lateral support must also be provided to comply with the detail in Fig. 3.19. Anchors should be of galvanised mild steel, 30 mm × 5 mm and spaced at 2 m maximum in houses not exceeding three storeys and 1.25 m maximum in all other buildings up to six storeys.

The effective height of a column is expressed as the distance between lateral supports or twice that distance where lateral support is not provided, as shown in Fig. 3.20. Piers are considered in the

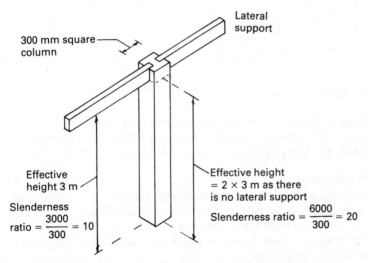

Lateral support

300 mm square column

Effective height 3 m

Slenderness ratio $= \dfrac{3000}{300} = 10$

Effective height $= 2 \times 3$ m as there is no lateral support

Slenderness ratio $= \dfrac{6000}{300} = 20$

Fig. 3.20 Example showing effective height and slenderness ratio relationship to columns

same way as walls unless their thickness is over 1.5 times the thickness of the wall of which they form a part. In this case piers are treated as columns in a right-angular plane to the wall.

Effective length

This is either:

1. Three-quarters of the clear distance between vertical lateral supports or twice the distance between a support and free edge where enhanced vertical resistance to lateral movement is provided (see Fig. 3.21); or

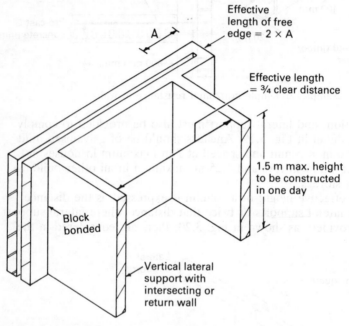

Fig. 3.21 Enhanced vertical resistance

2. The clear distance between vertical lateral supports or $2\frac{1}{2}$ times the distance between a support and a free edge where simple vertical resistance is provided (see Fig. 3.22).

Effective thickness

Effective thickness of single leaf walls and columns not stiffened by piers or intersecting walls is taken as the actual thickness. For cavity walls the effective thickness is taken as two thirds of the sum of the actual thickness of both leaves or the actual thickness of the thicker leaf, whichever is greater.

The effective thickness of walls stiffened by piers or intersecting walls is less easily defined. It is based on the formula:

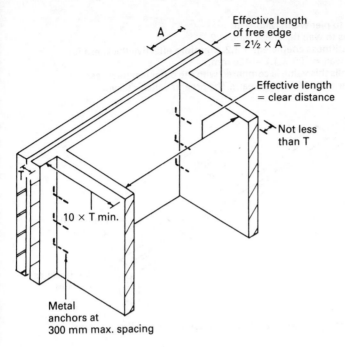

Fig. 3.22 Simple vertical resistance

Effective thickness $= t \times K$
where $t =$ actual wall or leaf thickness
and $K =$ stiffness coefficient.

Stiffness coefficient is based on pier spacing and K is found from Table 3.1. Interpolation is permitted.

Table 3.1

Pier spacing (centre to centre) to pier width ratio	Ratio of pier thickness to actual wall thickness to which it is bonded		
	1	2	3
6	1	1.4	2
10	1	1.2	1.4
20	1	1	1

An example to calculate effective thickness with the use of this table is shown in Fig. 3.23.

Further requirements for cavity walls

Cavity walls must be securely tied together with wall ties conforming to BS 1243, and be spaced in accordance with Building Regulation

52

Pier spacing to pier width ratio = 10
Pier thickness to wall thickness ratio = 2
From table, stiffness coefficient = 1.2, therefore effective thickness for
a single leaf wall = 100 × 1.2 = 120 mm.
For cavity walls this value is compared with the outer leaf thickness
and 2/3 (outerleaf + [K × inner leaf]). The greater value is taken.

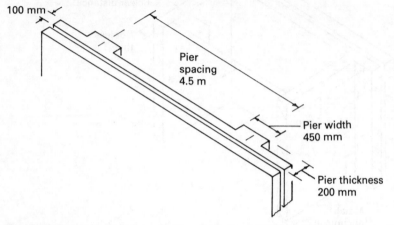

Fig. 3.23 Example showing effective thickness of walls with piers

requirements. The cavity may vary between 50 and 150 mm but
should never exceed 75 mm where either leaf is less than 90 mm in
thickness. With all spacings the ties must embed in at least 50 mm of
mortar in each leaf. For normal house construction the differential
movement between outer and inner leaf will not be significant
enough to loosen wall ties, but for larger buildings the outer leaf
should be supported at intervals of every third storey or 9 m,
whichever is less. A four-storey building not exceeding 12 m in
height is excepted and may have uninterrupted brick outer leaf
cladding for its full height. Vertical joints to absorb horizontal
movement are required where the wall exceeds about 12 m in
length. These are provided at 12 m intervals to accommodate
expansion of the brickwork by leaving a vertical gap of not less than

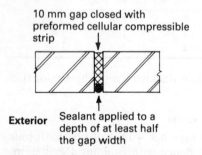

Fig. 3.24 Movement joint in brickwork

10 mm between adjacent walls. The detail in Fig. 3.24 shows suitable treatment in this situation. Back-up materials most suited for expansion joints include cellular polyethylene, cellular polyurethane, expanded rubber, sponge rubber and preformed mastic strip. Polysulphide and silicon sealants are preferred as these are not bitumenous or oily, but selection should be with consultation with the manufacturers of the back-up material.

Reinforced concrete and steel-framed buildings

Reinforced concrete

The following section is not intended as a structural design guide, but as an introduction to some of the factors determining structural stability. For a full appreciation it is necessary to refer to British Standard, Codes of Practice:

CP 110 : Part 1 : 1972, *The Structural Use of Concrete*

CP 111 : 1970, *Structural Recommendations for Loadbearing Walls.*

CP 114 : Part 2 : 1969, *Reinforced Concrete in Buildings.*

Plain concrete walls

Concrete walls are designed on the same principles as unreinforced brick or block masonry walls.

If the length of the cast *in situ* wall exceeds 2 m the likelihood of cracking by shrinkage and temperature movement is high. The effect is unlikely to have any structural significance but its control is necessary to preserve the finish or surface treatment. Steel reinforcement is the most practical method of containing cracks and the amount used should be no less than 0.3 per cent mild steel or 0.25 per cent high-yield steel. The percentage is expressed as a proportion of the total volume of concrete in the wall. Reinforcement is by small diameter bars, closely spaced and dispersed in the ratio of two thirds horizontal and one third vertical. For optimum effectiveness it must be close to the surface with regard to the quality of concrete and degree of exposure. In no case may it be less than 15 mm. Internal walls are more likely to crack at intersections and junctions with floors and beams, therefore reinforcement is concentrated in these areas.

Openings in both internal and external walls are provided with nominal reinforcement as shown in Fig. 3.25. These are areas of weakness with concentrated loads in the surrounding concrete. For calculation of effective height, a maximum factor of 30 times the wall thickness is permitted. In an unbraced wall the effective height is taken as one and a half times the wall height where horizontal restraint is provided by a floor or roof. Otherwise it is taken as twice the wall height with the exception of gable walls where the height is measured half way up the gable. Effective height factors are represented in

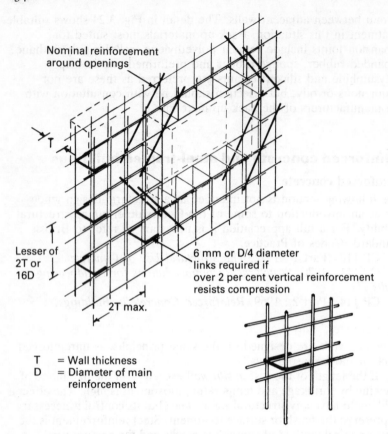

Fig. 3.25 Support and arrangement of reinforcement in a concrete wall

Fig. 3.26 with details of the effect where lateral support provides resistance to movement and rotation. If provision for absorbing movement due to expansion and contraction is required, suitable vertical joints should be provided at 15 m maximum spacing.

Reinforced concrete walls

The minimum requirements for a concrete wall to classify as reinforced are illustrated in Fig. 3.27. For most purposes effective height is calculated in the same way as plain concrete walls. However, better results are obtained if the wall is constructed monolithically with adjacent construction.

Due to the effect of reinforcement, slenderness ratios are higher than masonry walls or plain concrete walls. If the wall is braced and reinforced the slenderness ratio should not exceed 40, unless over 1 per cent of the horizontal section area is vertically reinforced which

55

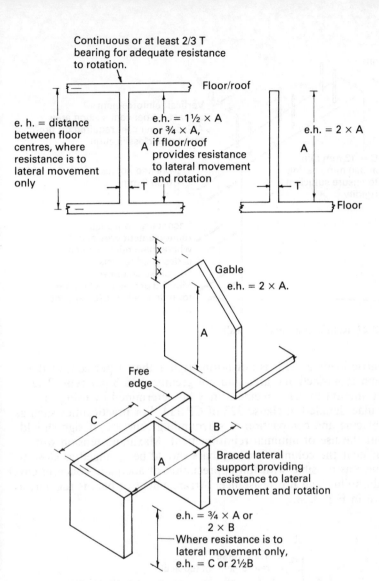

Fig. 3.26 Effective height of concrete walls

permits an increase to 45. Unbraced walls have a slenderness ratio not exceeding 30.

Columns

Column sections are generally designed uniformly square, round or octagonal so that the radius of gyration is equal about both axes. Reinforcement is provided to resist tensile stresses caused by

56

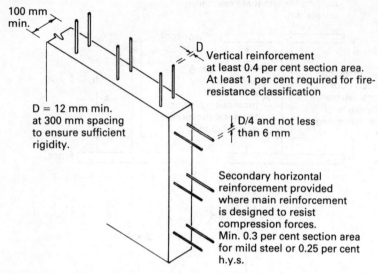

100 mm min.

D

Vertical reinforcement at least 0.4 per cent section area. At least 1 per cent required for fire-resistance classification

D = 12 mm min. at 300 mm spacing to ensure sufficient rigidity.

= D/4 and not less than 6 mm

Secondary horizontal reinforcement provided where main reinforcement is designed to resist compression forces. Min. 0.3 per cent section area for mild steel or 0.25 per cent h.y.s.

Fig. 3.27 Reinforced concrete wall

eccentric loading. This is not normally less than 1 per cent of the column cross-sectional area and not greater than 8 per cent. The exact amount of reinforcement may be determined by using the formulae detailed in clause 322 of CP 114. For practicalities such as the placing and compaction of surrounding concrete, design should favour the use of minimal reinforcement. Maximum tension will occur near the column face and bars should be arranged as close to the face as possible with full appreciation of adequate concrete cover for durability and fire resistance. Arrangement of bars is regular, as shown in Fig. 3.28.

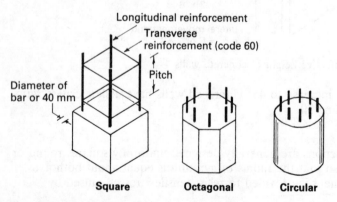

Longitudinal reinforcement

Transverse reinforcement (code 60)

Pitch

Diameter of bar or 40 mm

Square   **Octagonal**   **Circular**

Fig. 3.28 Regular reinforced concrete columns

Lateral or transverse reinforcement is provided to prevent buckling of longitudinal reinforcement and subsequent spalling of concrete in the immediate area. The diameter should never be less than one quarter of the main reinforcement and in no case less than 6 mm. The transverse reinforcement, more popularly termed links, should not be over spaced. The recommended pitch is no more than the least of:

(a) the least lateral dimension of the column,
(b) twelve times the diameter of the smallest longitudinal reinforcement,
(c) 300 mm.

For structural design it is necessary to define a column as either short or slender. Figure 3.29 illustrates the factors distinguishing the two. A short column has an effective height to both major and minor axis dimensions b and h, of less than 12. Otherwise it is considered slender up to a maximum of 60. Effective height varies between 0.75 × height and 2 × height, depending on the efficiency of directional restraint and bracing. Guidance is provided in Table 15 of CP 110.

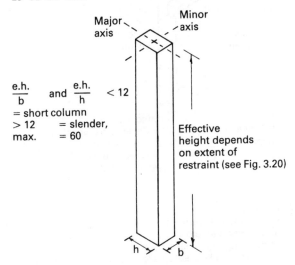

Fig. 3.29 Slenderness and effective height of columns

Beams and slabs

For most design purposes, beams and slabs are considered the same. Slabs are regarded as a series of beams 1 m wide, laid parallel to the span. If the slab is supported on four sides, the span is assumed to be the shorter direction unless the ratio of the spans is less than two. Design must then be in accordance with clause 314 of CP 114.

Whilst it is unnecessary to consider the details of reinforced concrete beam design in this subject at this level, it is nevertheless worth considering some of the construction principles contributing to stability and sound practice.

Effective span and effective depth. These are shown in Fig. 3.30, effective span taken as the distance between centre of bearing or the clear distance between support plus the effective depth, whichever is less.

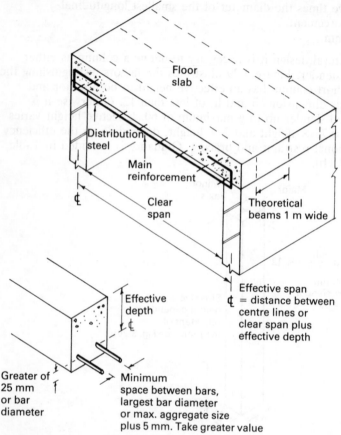

Fig. 3.30 Effective span and effective depth

Minimum reinforcement in slabs:

(a) Main steel 10 mm diameter.
(b) Distribution steel 6 mm diameter.
(c) The reinforcement in each direction should be no less than 0.15 per cent of the cross-sectional area for plain steel bars or 0.12 per cent for high-yield steel.

An approximate estimate of slab thickness can be assumed to be 40 mm for every 1 m span, with 100 mm accepted as a minimum. For calculation of the minimum amount of tensile reinforcement in a beam, reference must be made to clause 306 in CP 114, where the use of simplified formulae is illustrated.

Stiffness and slenderness. Reinforced concrete must have adequate stiffness to prevent deflection or deformation which might impair the strength or affect surface finishes. The maximum acceptable deflection is 1/250 of the basic span and this will normally resist damage to finishes up to a maximum span of 10 m. The basic span to effective depth ratios containing this maximum deflection are shown in Fig. 3.31.

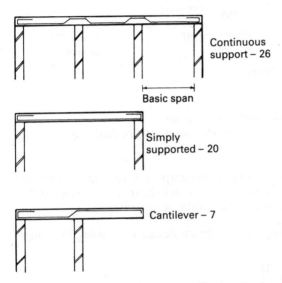

Continuous support – 26

Basic span

Simply supported – 20

Cantilever – 7

Fig. 3.31 Maximum basic span to effective depth ratios

The slenderness of a beam is affected by breadth and depth ratio and the arrangement of lateral support. The clear distance between supports must not exceed 60 times the breadth of the compression face of a beam or 250 times the breadth of the compression face squared, divided by the effective depth. The least value is applied. A simplified explanation is shown in Fig. 3.32.

Steel-framed buildings

The design of steel-framed structures is a complex subject and this section undertakes to provide no more than an appreciation of the principles applied to simply loaded beam and column calculations. For a complete understanding reference should be made to:

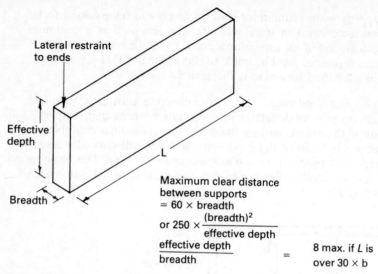

Fig. 3.32 Slenderness of beams

BS 4 : Pt 1 : 1980, *Structural steel sections.*
BS 449 : Pt 2 : 1969, *The use of structural steel in building.*

Simple beam design
With structural steel beams the main purpose is to achieve a section capable of resisting bending moments and shear forces, while still providing adequate stiffness and resistance to deflection.

Bending. The initial approach to beam design is by using the simple formula:

$$Z = \frac{M}{f}$$

where $Z =$ section or elastic modulus (denoted in BS 4),
$M =$ moment of resistance which must at least equal the maximum bending moment,
$f =$ fibre stress of the steel, (165 N/mm^2 for rolled I beams)

The bending moment at any one point on a beam is established by subtracting the load moment from the reaction moment. For simple supported beams of uniform loading, such as those shown in Fig. 3.33, the bending moment is easily calculated.

Shear
Shear forces in a simply supported beam are maximum at the support points, but this is not usually critical as design to withstand the bending moment will normally be more than sufficient for shear

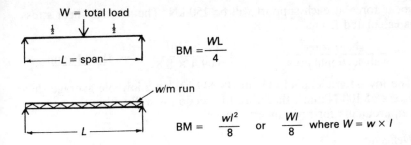

$$BM = \frac{WL}{4}$$

$$BM = \frac{wl^2}{8} \quad \text{or} \quad \frac{Wl}{8} \quad \text{where } W = w \times l$$

e.g. a simply supported beam of span 5 m carrying a udl of 300 kN.

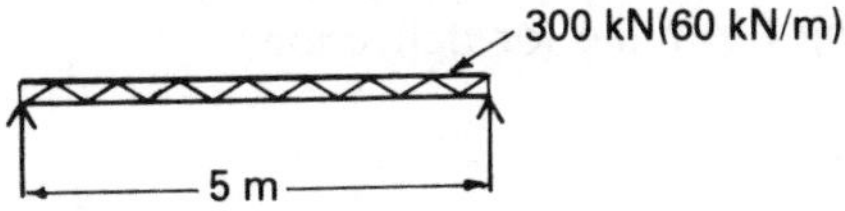

300 kN(60 kN/m)

5 m

$$BM = \frac{wl^2}{8} = \frac{60 \times 5^2}{8} = 187.5 \text{ kNm}$$

$$Z = \frac{M}{f} = \frac{187.5 \times 10^6}{165} = 1136.36 \text{ cm}^3$$

Reference to BS 4, Table 5, UBs indicates that a $406 \times 178 \times 67$ kg/m section of elastic modulus about the X–X axis of 1188 cm³ will satisfy these requirements.

Fig. 3.33 Bending moment calculation in simply supported beams

Serial size $406 \times 178 \times 67$ Kg/m

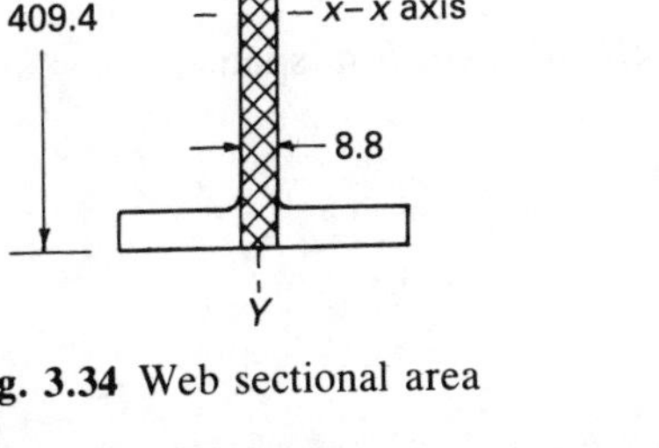

Y

Web sectional area

409.4 — — x–x axis

8.8

Y

Fig. 3.34 Web sectional area

purposes. Calculation is based on the average stress value acting on the web sectional area, as shown in Fig. 3.34. Using the figure of 300 kN distributed load from the previous example, the maximum

62

shear force at each support will be 150 kN. The average shear stress is calculated from:

$$\frac{\text{shear force}}{\text{web sectional area}} = \frac{150 \times 10^3}{409.4 \times 8.8} = \underline{41.63 \text{ N/mm}^2}$$

The lowest grade steel (43) in BS 449 has an allowable average shear stress of 100 N/mm², therefore the section is well within the requirements for resistance to shear.

Deflection

Deflection is specified in BS 449 as a maximum of 1/360 of the span. It must not impair the strength and efficiency of the structure or damage surface finishes and is attributed only to imposed load, not structural load. Figure 3.35 shows two simply-supported beam situations, with the corresponding formulae for deflection.

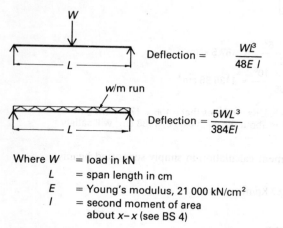

Where W = load in kN
 L = span length in cm
 E = Young's modulus, 21 000 kN/cm²
 I = second moment of area about x–x (see BS 4)

Fig. 3.35 Deflection of simply supported beams

Example

Using the previous figure of 300 kN *udl* over 5 m span.

$$d = \frac{5}{384} \quad \frac{Wl^3}{EI}$$

$$= \frac{5 \times 300 \times 5^3 \times 100^3}{384 \times 21\,000 \times 24\,329}$$

$$= 0.95 \text{ cm}$$

Permissible deflection = $1/360 \times 500 \text{ cm} = \underline{1.39 \text{ cm}}$

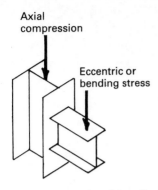

Fig. 3.36 Axial and bending stresses

Columns or stanchions

Columns are subject to both axial compression stresses and bending or eccentric stresses. These are represented in Fig. 3.36. In practice, eccentric stresses are more likely and this is accounted for further on. For purposes of stanchion design subject to compressive stresses only, two factors must be considered:

1. Length of member.
2. Manner in which the ends are secured.

Long compression members have a tendency to buckle or produce side bending under extreme load. This is partially attributed to the length, cross-sectional area and end fixing but the shape of section will have a more significant effect. The cross-section dimensions enter into column calculations by using a geometrical property of section called the radius of gyration. It is a dimension which is defined as the square root of the second moment of area, divided by the cross-sectional area.

$$r = \left(\frac{I}{A}\right)^{1/2}$$

Fortunately this is calculated and tabulated in BS 4. It is important to appreciate that a column will normally buckle about its weaker axis, therefore the radius of gyration about the y–y axis is used for calculation.

Effective length and slenderness ratio

Where the ends of a column are located at specific points by beams or other means of retention, they are regarded as fixed in position and direction, but if the ends are hinged or pinned they are position-fixed only. Figure 3.37 illustrates the possibilities and the effect on column length. The application of effective length and slenderness

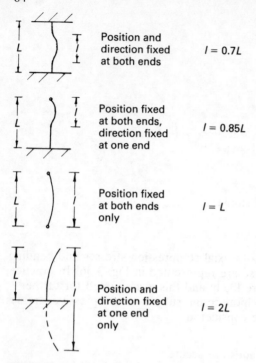

Position and direction fixed at both ends $l = 0.7L$

Position fixed at both ends, direction fixed at one end $l = 0.85L$

Position fixed at both ends only $l = L$

Position and direction fixed at one end only $l = 2L$

Fig. 3.37 Effective length 'l' of compression members

ratio are demonstrated in the following simple examples of column design:

Determine the maximum axial load a UC, $203 \times 203 \times 46$ kg/m of length 8 m can carry if it is position-fixed at both ends but restrained in direction at one end only.

Effective length, $l = 0.85 \times 8 = 6.8$ m.
Radius of gyration (r) about the y–y axis from BS 4 $= 51.1$ mm
Slenderness ratio, $\dfrac{l}{r} = \dfrac{6.8 \times 10^3}{51.1} = \underline{133}$

Reference to Table 17(a) in BS 449 provides a maximum allowable stress of 50 N/mm², therefore total axial load is 50 multiplied by the cross-section area (from BS 4),

$$= \frac{50 \times 58.8 \times 10^2}{10^3} = \underline{294 \text{ kN}}$$

Design to accommodate eccentric loads involves determining whether a section will withstand a combination of the anticipated axial stress and the bending stress. A stanchion must not carry an axial load in excess of maximum allowable stress (Table 17(a) BS 449), therefore,

$$\frac{\text{axial load } (fc)}{\text{allowable axial compressive stress } (pc)} = \text{not greater than 1}$$

When subject to bending,

$$\frac{\text{maximum compressive stress } (fbc)}{\text{allowable bending stress (Table 3(a) BS 449) } (pbc)} = \frac{\text{not greater}}{\text{than 1}}$$

Hence when subject to both axial and bending compression, the amount of axial compression stress added to the bending stress must also not exceed unity, thus,

$$\frac{fc}{pc} + \frac{fbc}{pbc} = \text{not greater than 1}$$

The following example in Fig. 3.38 shows these effects.

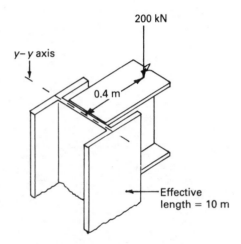

Fig. 3.38 Effect of axial and compressive loads

Try 254 × 254 × 107 kg/m UC.

$$\frac{l}{r} = \frac{1000}{6.57} \text{(from BS 4)} = \underline{152}$$

Reference to Table 17(a) BS 449, $pc = \underline{39}$ N/mm^2

$$\frac{D}{T} \text{ (see BS 4)} = \underline{13}$$

Reference to Table 3(a) BS 449, $pbc = \underline{150}$ N/mm^2

$$fc = \frac{\text{Load}}{\text{area}} = \frac{200 \times 10^3}{136.6 \times 10^2 \text{(BS 4)}} = \underline{14.65} \text{ N/mm}^2$$

$$fbc = \frac{\text{Bending moment}}{\text{Z (elastic modulus about } X-X)} = \frac{200 \times 400}{1313} = \underline{61} \text{ N/mm}^2$$

hence, $\quad \dfrac{fc}{pc} + \dfrac{fbc}{pbc} = \dfrac{14.65}{39} + \dfrac{61}{150} = \underline{0.78}$

As this figure is less than 1, the section is satisfactory.

Chapter 4

Foundations, factors affecting design and suitability

This chapter is a continuation of the information on raft, strip and short-bored piled foundations considered in Chapter 5 of *Construction Processes, Level 1*. This prerequisite reading is necessary to appreciate the effect that ground heave and shinkage and building settlement can have on the structure. This chapter considers these effects in a little more detail and continues to outline the use of column foundations in difficult situations.

Soil properties

Soils need classification in order to understand their properties and the triangular chart provides a useful classification for subsoils of predominantly sand, clay and silt composition. This is shown in Fig. 4.1, identifying a clay loam subsoil composed of 40 per cent sand, 35 per cent silt and 25 per cent clay.

British Standard 1377 : 1975, *Methods of Test For Soil For Civil Engineering Purposes*, uses particle size as a means of grouping soils.

These are:

Clay, particles up to 0.002 m.
Silt, particles between 0.002 and 0.06 m.
Sand, particles between 0.06 and 2 mm.
Gravel, particles between 2 and 60 mm.
Cobbles, particles between 60 and 200 mm.

68

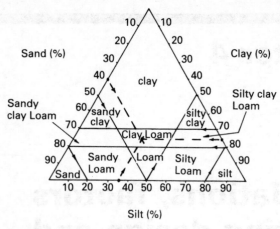

Fig. 4.1 Triangular chart

Combined with these particles will be found water and, to a lesser extent, air. The water content is responsible for the cohesive nature and strength of fine particled soils, whilst larger particled soils rely on their weight and closeness of packing for strength. Thus, large particles have little strength when loose and water will have an insignificant effect.

When small-particled soils dry due to the effect of dehydration from drought or the effect of tree roots, the solid particles are drawn together. Sands with little water are hardly affected but clays will shrink, becoming materially stronger in content although generally deteriorating because of the creation of voids and fissures.

Soils are subject to extreme pressure in the area immediately below a foundation and consolidation increases both water and soil pressures. The water content disperses while the soil particles are compressed closer together until the forces between them equal the new loading. The effect on sandy subsoils is unnoticed or appears negligible as settlement by consolidation is a quick process occurring during the building construction. Clay subsoils are less responsive as the water content is more difficult to disperse, and settlement can continue for several years. However, even in subsoil of uniform composition the amount of settlement is difficult to ascertain. It will depend on the applied pressure and the property of the soil contained within the geometric envelopes shown in Figs 4.2, 4.3 and 4.4.

The bulb distribution of shear stress under a strip foundation shows that maximum shear occurs at points around the semicircle of diameter equal to the foundation breadth. The prismoidal diagram provides an indication of the pressure exerted at descending levels of subsoil. It provides no more than an approximate value, useful for

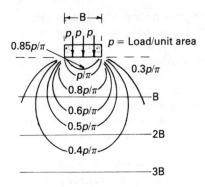

Fig. 4.2 Bulb of shear stress

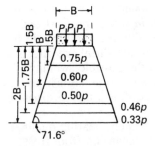

Fig. 4.3 Prismoidal pressure diagram

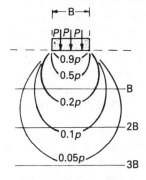

Fig. 4.4 Bulb of pressure

obtaining preliminary information about the effects of vertical loading. The bulb of pressure is more accurate, allowing for the fact that loaded subsoils react to both vertical and horizontal pressure with circular stress levels extending through different depths. With isolated pad foundations to columns, the geometric form will be globular about the column axis, whereas a strip foundation will produce a bulb shape which is cylindrical along the foundation length.

Subsoil movement may also be attributed to climatic changes, the effects of vegetation and ground subsidence. Clay soils which shrink and swell to any extent will cause movement to shallow foundations. Some clays are sufficiently firm to support low rise buildings without excessive movement, but any shrinkable clay is suspect and on open sites, away from the influence of trees, a 900 mm-deep reinforced concrete strip foundation must be considered. To economise in materials and to provide a satisfactory balance between building load and subsoil-bearing capacity, a narrow trench of about 400 mm is provided. The trench is cut by mechanical excavator to a high degree of accuracy, otherwise brick footing courses will project over the edge of the concrete filling.

Buildings to be constructed on clayey soiled sites with trees are best provided with piled foundations. Cast *in situ* or pre-cast end-bearing piles may be chosen to suit site conditions. They should be sufficiently deep enough to be immune from the shrinkage effects caused by trees drawing ground water and the possible swelling following tree felling. Illustrations of deep strip foundations, pre-cast and *in situ* short-bored piled foundations are shown in *Construction Processes, Level 1*.

Sandy soils, once consolidated, offer excellent bearing qualities, unless affected by running ground water or expansion from frost. Underground water flow causes loss of ground by washing out the finer particles, leaving the coarser material in unstable condition. Sandy subsoiled sites having a high water table should be avoided for building purposes unless they are provided with a permanent, efficient means of subsoil drainage. Frost heave is possible with fine sand, silt and chalk. With these subsoil materials, water moving into the frozen area freezes between the particles, increases in volume and lifts the ground surface. There is greater tendency for slabs or raft foundations to lift as these are close to the surface. Strip foundations constructed at least 600 mm below the surface are unlikely to be affected.

Larger scale movement is possible from landslip, particularly if the subsoil material is clay and the land slope exceeds about 1 in 10. Swallow holes and mining subsidence have a similar effect, characterised by the overburden falling and leaving a large recess in the surface. The former is a natural geological failure caused by the creation of large cavities in chalk and limestone subsoils by the erosive effects of underground water. Mining subsidence is of course caused by an artificial or man-made void which may have existed, undetected and unrecorded, for several centuries.

Foundations to reinforced concrete-framed buildings

Independent pad foundations to columns

Pad foundations are square or occasionally rectangular where uniform distribution of concrete is restricted. The area of concrete is ample so as not to overstress the subsoil below. Reinforcement is necessary to resist bending, and this will normally be sufficient for resistance to shear failure also. The critical positions for shear and bending are indicated in Fig. 4.5, with a typical detail of the construction of a reinforced concrete pad foundation.

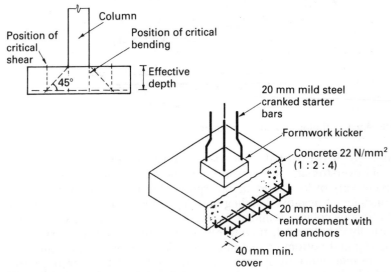

Fig. 4.5 Independent pad foundation

Continuous column foundation

A continuous column foundation is a form of strip foundation used where columns are very closely spaced and separation of pads is uneconomical. It could also be used where the spread of pad foundations square to the line of columns is restricted by a boundary or other building. The strip is designed as a continuous reinforced beam with the columns as point loads and the subsoil as a uniform resistance. A diagrammatic representation and section showing allocation of main reinforcement are shown in Fig. 4.6.

Combined column foundation

This is used where there is an insufficient foundation bearing area between the column positions and a line of restriction. The perimeter columns combine foundations with an internal line of

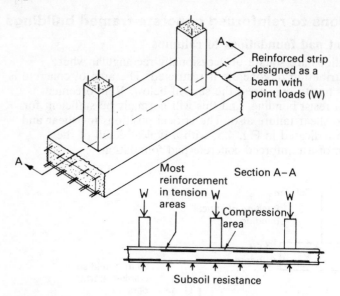

Fig. 4.6 Continuous column foundation

columns. It is essential to provide a sufficient pad bearing area to avoid overstressing the subsoil at any point, and in particular any eccentricity of load must be incorporated in the pad design. The principles of negative bending between columns are the same as for continuous column foundations except the outer column support is partially cantilevered, therefore tensile reinforcement will extend to the top and bottom of the outer pad. Furthermore, the column centre lines must always coincide with the foundation centre of area, and where column loads differ or a space restriction exists next to the heaviest column, a trapezoidal combined pad foundation may be used to accommodate load differences. Figure 4.7 illustrates these principles and emphasises the coincidence of column centres of gravity with a foundation central area.

Cantilevered column foundation

Cantilevered foundations are required where the perimeter structure must be suspended over an obstruction. This is to avoid pressurising the area below a column which may form part of an existing substructure or contain services which cannot be moved. The column is supported by a ground beam which cantilevers beyond a fulcrum base, positioned on a positive location within the structural frame. Counterbalancing is most conveniently provided by an internal column designed to carry a load well in excess of the suspended column in order to provide an adequate safety margin. If sufficient load cannot be obtained from an adjacent column, mass concrete

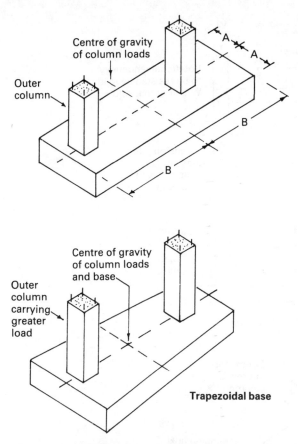

Fig. 4.7 Combined column foundation

may be provided as a counterbalance or tension piles could be used as ground anchorage. An alternative application uses a large slab base to provide an increased bearing area. Here the cantilevered ground beam bears directly on the slab before extending out to receive a perimeter column. See Fig. 4.8.

Balanced base foundation

A balanced base foundation can be used where limited space prevents concentric location of column and base. To prevent rotation, a beam links the outer base to an inner concentrically loaded base. The balancing beam could have an eccentric load distribution effect on the inner column base, and to avoid rotation here, the base must have sufficient ground area to carry the column loads safely and to prevent overstressing of the subsoil near its far edge. See Fig. 4.9.

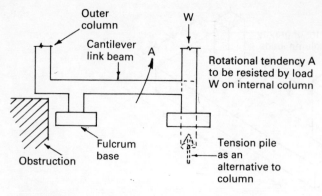

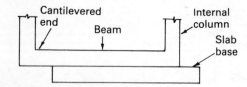

Fig. 4.8 Cantilevered foundation

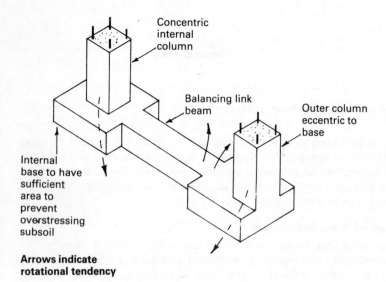

Fig. 4.9 Balanced base foundation

Chapter 5

Industralised and rationalised building techniques

The application of industrialised and rationalised terminology to the construction process indicates the continuing trend into factory manufacture and assembly of building elements. The rational aspects include alternative, more economical use of materials, providing a considerable reduction in the number of skilled and semi-skilled operatives employed on the building site.

This chapter considers further the design concepts of timber-framed housing and provides an introduction to the use of pre-cast concrete components in housing and commercial buildings.

Timber-framed house construction

Timber-framed housing in the UK has developed from experiences gained in Scandinavia and North America. Variation in construction methods exist, although the principles are very similar. The three most popular methods are:

1. 'Stick-built'.
2. Balloon frame.
3. Platform frame.

'Stick' building is a Canadian technique involving site assembly of timber frames. Studs (sticks) are arranged vertically at between 400 and 600 mm spacing, with an internal arrangement of diagonal bracing or struts for stability. The stud framework is clad internally and externally with suitable lining materials.

UK timber-framed construction techniques are more prefabrication orientated, with stud walls factory assembled with exterior lining attached. Interior lining of plasterboard is not usually prefixed, to allow for site installation of electrical and piped services. The preparation of balloon and platform frame components are very similar, but site assembly varies, as shown in Fig. 5.1. Balloon framing consists of two-storey height panels installed in one operation, with the first-floor structure suspended from the panel stud framework. This has proved less suitable than platform framing, which has wall panels produced in single-storey heights to provide a stage for intermediate floors and upper wall panels. This permits easier prefabrication, transport and site assembly with reduced possibility of cracking to interior finishes caused by structural shrinkage and movement.

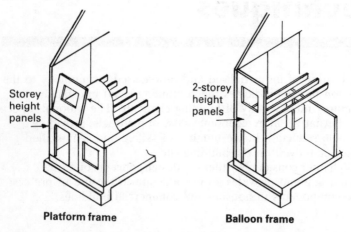

Storey height panels

2-storey height panels

Platform frame   **Balloon frame**

Fig. 5.1 Platform and balloon, timber frame construction

The following illustrations of timber-framed construction details supplement those shown in Chapter 1, and provide information on assembly of platform frame construction.

Methods of fastening and assembly

Fastening is by staples, nails and bolts. Bolts are rarely used, possibly only for securing sole plates to the ground floor slab or for joining adjacent panels. Nailing is the simplest and quickest method of joining timber to timber, and the majority of connections between structural components are made with 75 mm round wire nails at 300 to 600 mm spacing, depending on the situation. Pneumatically driven staples are used for factory assembly of sheathing and lining

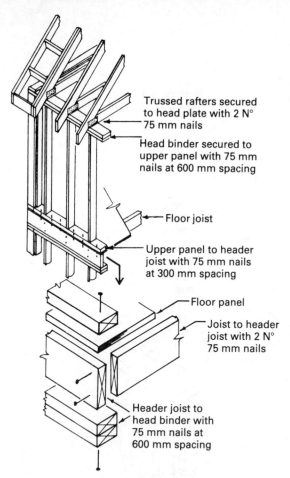

Trussed rafters secured to head plate with 2 N° 75 mm nails

Head binder secured to upper panel with 75 mm nails at 600 mm spacing

← Floor joist

Upper panel to header joist with 75 mm nails at 300 mm spacing

Floor panel

Joist to header joist with 2 N° 75 mm nails

Header joist to head binder with 75 mm nails at 600 mm spacing

Fig. 5.2 Assembly at first floor and eaves

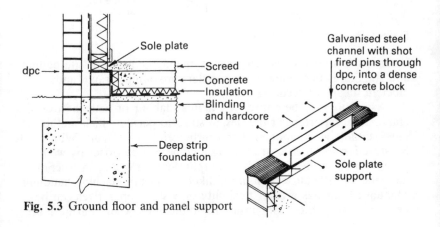

Sole plate

dpc →

Screed

Concrete

Insulation

Blinding and hardcore

Deep strip foundation

Galvanised steel channel with shot fired pins through dpc, into a dense concrete block

Sole plate support

Fig. 5.3 Ground floor and panel support

78

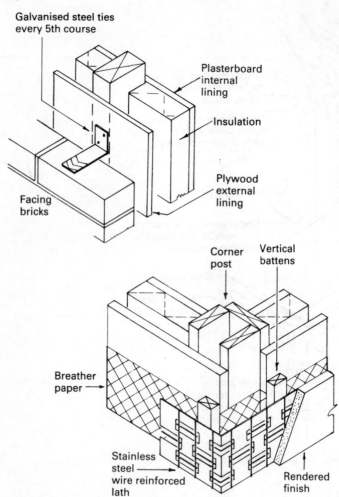

Fig. 5.4 External wall construction

materials to stud frames. The use of pneumatic guns for site nailing and stapling is becoming more popular, providing a rapid and efficient means of securing parts.

Assembly may be entirely by manual labour or with the use of a crane. It is important at the design stage to decide which method is most suitable as component size can be produced to suit. With estate construction the high cost of using a crane can be offset by well-programmed application and full output. Smaller scale construction and sites of difficult terrain are unlikely to justify machine handling. Manual labour has greater flexibility of cost, as operatives can be deployed elsewhere if organisational difficulties occur. Furthermore,

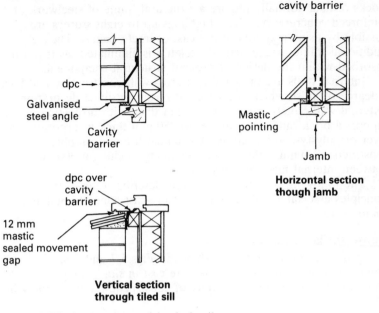

Fig. 5.5 Window section and jamb detail

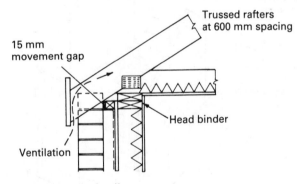

Fig. 5.6 Eaves detail

panelled components will require lifting eyes and possibly temporary bracing to counter the effect of lifting stresses.

Pre-cast concrete walls and floors

Construction with these units is suitable for all types of residential building as pre-cast concrete walls and floors adequately substitute for traditional brick and timber as structural components. High rise blocks of flats normally require a structural frame of steelwork or reinforced concrete; however, buildings up to eight storeys are possible by using a system of pre-cast concrete panels. The traditional frame as a structural skeleton is eliminated, as it can be absorbed within the design of pre-cast structural components.

Internal walls are principally of cross-wall construction and have a degree of finish which is capable of receiving direct decoration. External wall panels are either 100 mm thick with projecting ties to support a brick outer leaf, or 200 to 250 mm thick incorporating a layer of polystyrene insulation. To standardise and simplify construction, manufacturers use a planning module of 300 mm between internal faces of walls.

The following constructional forms describe some of the principles associated with assembly of pre-cast concrete building components.

House shells

The principles of assembly are shown in Figs 5.7 and 5.8. Foundations and ground floor slab are cast *in situ* usually monolithically to ensure continuity of construction and economy in

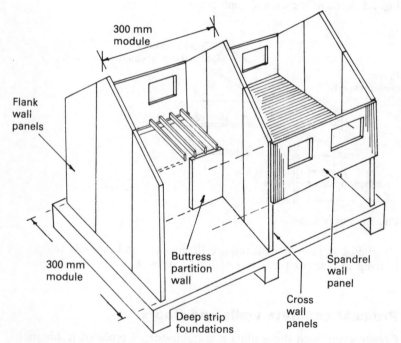

Fig. 5.7 Pre-cast concrete house shell construction

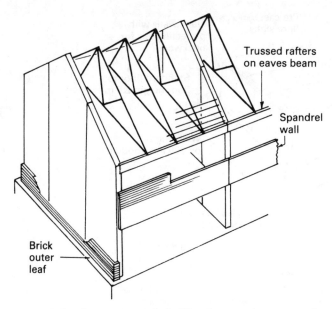

Fig. 5.8 Roof structure and cladding to pre-cast concrete house shells

labour. Flank walls and cross walls are erected and located on the prepared foundation, whilst partition and buttress walls are used to provide stability and to carry the upper floor joists. Front and rear panel walls may be formed by conventional cavity brick and block construction. This is likely to occur between ground and first floor with brick cladding to the complete height of flank walls. The upper floor wall is frequently constructed as a pre-cast concrete spandrel panel, which is unable to carry the roof load. Here an eaves beam is used to transmit roof loading to flank and cross walls.

Cross-wall

The major advantage of pre-cast concrete cross-wall construction is ease and speed of assembly of party and flank walls, plus compatibility with pre-cast concrete floor units. Additionally the cladding and face walls may be brick to preserve traditional features, although they are not expected to carry load. This is entirely the function of the pre-cast concrete internal panels. Figures 5.9 and 5.10 illustrate the principles and method of construction and Fig. 5.11 provides construction details at significant junctions.

Figure 5.9 shows *in situ* concrete foundations and ground floor slab with flank and cross-walls positioned to receive brickwork and upper floor. Pre-cast concrete floors have a finished upper and lower surface ready to receive floor tiles or carpet and the soffit, a textured paint. Figure 5.10 shows a repeat at first floor with brick exterior

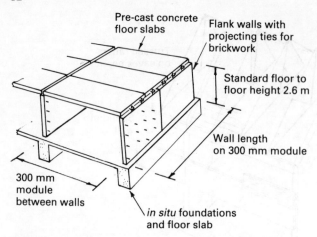

Fig. 5.9 Pre-cast concrete cross-wall construction

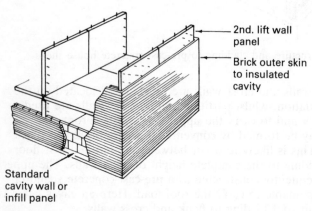

Fig. 5.10 Brick cladding to pre-cast concrete cross-wall construction

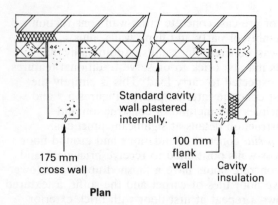

Plan

Fig. 5.11 Construction at pre-cast concrete flank and cross-walls

cladding. This continues to the required height but direct support for the outer leaf of brickwork must be provided at lifts of 9 m maximum or three storeys. The front and rear panels have a standard cavity wall with inner leaf built between the loadbearing walls. Details are shown in Fig. 5.11. Alternatively, a complete prefabricated infill panel may be fitted between cross and flank walls. Figure 5.12 shows application to flats with standard pre-cast stair flights and the detail in Fig. 5.13 shows construction between flank walls. Flat roofs are provided with pre-cast floor/roof slabs or for pitched and tiled covering, trussed rafters are possible.

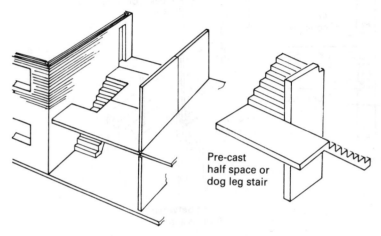

Pre-cast half space or dog leg stair

Fig. 5.12 Pre-cast concrete cross-walls, application to flats

Wall frame

This is a variation of cross-wall construction using complete pre-cast concrete cavity wall panels with sandwiched insulation. The assembly principles are similar, although the components are more suited to taller structures of flats or maisonettes. The external facing material is concrete, offering a wide choice of surface profile and texture. Construction details are shown in Figs 5.14 and 5.15.

Pre-cast framed building for multi-storey structures

This type of construction is suitable for commercial buildings, offices and laboratories. In similarity with other forms of industrialised building, it offers considerable saving in construction time compared with other methods. The design provides the largest possible area of enclosed space without the interference of intermediate structural components. If particularly large spans or openings are required, the system is sufficiently flexible to include supplementary columns and beams. These may also relieve the loading from external walls, providing the possibility of variation using infill panels or open

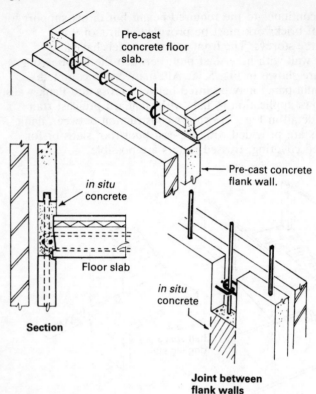

Fig. 5.13 Cross-wall construction details

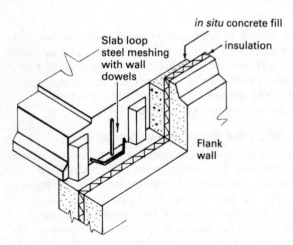

Fig. 5.14 Wall frame, construction between wall and floor

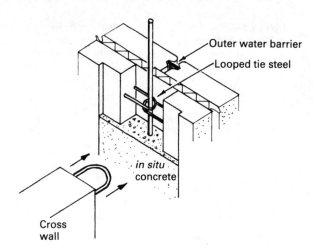

Fig. 5.15 Wall frame, construction between panel walls and cross wall

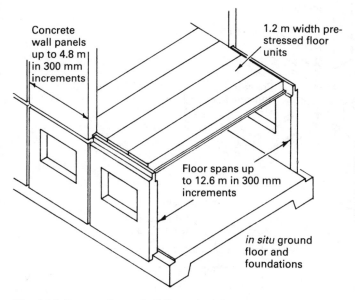

Fig. 5.16 Pre-cast frame building principles

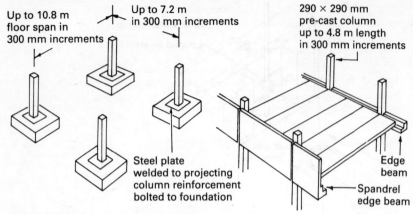

Fig. 5.17 Pre-cast frame with columns and beams

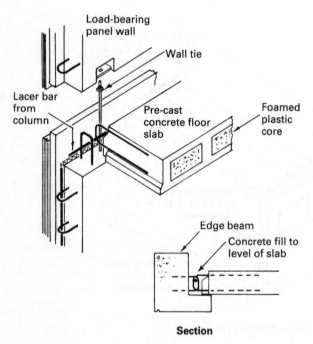

Section

Fig. 5.18 Pre-cast frame construction details

balconies. The preceding illustrations indicate the construction using loadbearing walls and columns and beams.

Pre-cast concrete floor systems

These are available in many forms, originating from the use of hollow clay block and *in situ* concrete composite floors shown in Fig. 5.19. Concrete floors are available in many forms, including plank, box and beam sections, all providing considerable saving in

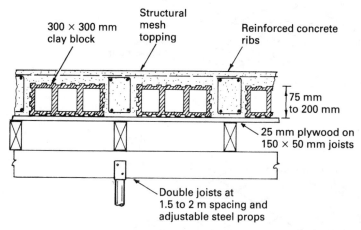

Fig. 5.19 Hollow clay block floor and temporary support

construction time over conventional poured *in situ* concrete. Further advantages include a considerable reduction in structural weight as most are hollow or very thin, properties provided by prestressing the concrete with reinforcement steel. They require no formwork and any units designed to receive a structural topping use the preformed unit as permanent formwork. Although formwork is omitted, some systems will require temporary props during construction. These are mainly composite systems having a superimposed structural topping and some of the wide slab floors which will also need props for levelling at mid-span. All systems are steel reinforced to resist tensile and shear stresses, and the majority are prestressed to economise in concrete and to permit large spans. The details in Fig. 5.20 provide examples of hollow pre-cast concrete floor/roof with screed and reinforced concrete topping, and Fig. 5.21 indicates the application to ground floors in housing, using prestressed concrete ribs and lightweight infill blocks.

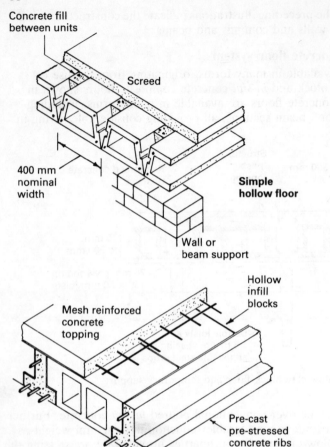

Fig. 5.20 Pre-cast concrete hollow floor construction

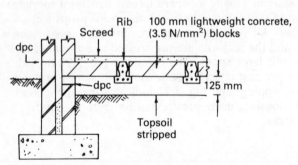

Fig. 5.21 Pre-cast concrete suspended ground floor.

Chapter 6

Internal services

Cold- and hot-water installations

Cold-water supply

The installation of pipework for the conveyance of cold water into and within a building must satisfy the requirements of the local water authority by-laws. These are independent of other building legislation and vary slightly between the different area authorities. Figure 6.1 shows the installation principles for a supply pipe from the water company's main below the road to the householder's stop valve inside the dwelling. The plug valve can be drilled and fitted to the existing cast-iron main under pressure, whilst a new main and supply pipe will be provided with a stop valve.

Snaking or a goose neck is provided to absorb settlement of the main and contraction of plastic pipework during warm weather. A minimum cover of 750 mm is required primarily to avoid frost damage and also the effects of vegetation and gardening. The communicating pipe is owned and maintained by the water authority and is usually terminated with a stop valve within the property boundary. This valve is for the householder to isolate the service pipe for maintenance and repair. It is also useful to shut off supply during periods when the house is unoccupied. Service pipe entry to the building must be protected against settlement of the structure. The clayware drain pipe in Fig. 6.1 sealed with a non-perishable absorbent filler at both ends is a possibility here. Alternatively a small concrete lintel could be provided over the service entry area.

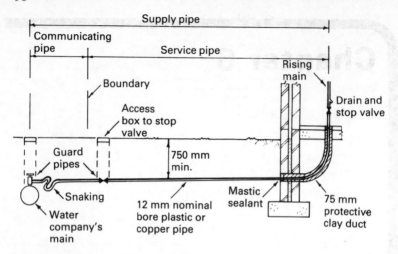

Fig. 6.1 Water supply to a building

Several pipe materials are viable for distribution services; plastic, galvanised steel, stainless steel and copper being some of the most popular. Since metrication, confusion has arisen due to manufacturers of some materials changing their dimensional specifications. As most domestic pipework installations are undertaken in copper tube the following dimensions will be used in the illustrations:

15 mm outside diameter, formerly $\frac{1}{2}$ in nominal bore.
22 mm outside diameter, replaces $\frac{3}{4}$ in nominal bore.
28 mm outside diameter, formerly 1 in nominal bore.

Distribution of cold water

Local authority by-laws will determine the arrangement of piped water services within a building. Where mains pressure can be maintained at a constant high value it is possible to supply all fittings direct. This is mainly limited to northern areas of Britain, where supply reservoirs can be situated at relatively high levels. In other areas an indirect supply system is more suitable. Here all fittings, with the exception of the kitchen sink tap, are supplied from a cold-water storage cistern. Both systems are represented by single line diagrams in Figs 6.2 and 6.3, where the simpler direct system indicates a more economic installation. The main advantages are the use of a smaller storage cistern (115 litres actual capacity) resulting in a cheaper installation and less dead load on the ceiling, and the provison of drinking water at all outlets. The disadvantages are that there are no reserves if the supply is interrupted and the pressure at outlets is variable, which is particularly noticeable during peak periods. Where pipework and fittings are badly installed this could

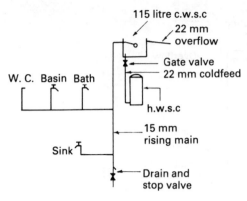

Fig. 6.2 Direct cold water supply

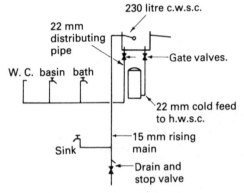

Fig. 6.3 Indirect cold-water supply

cause back siphonage, i.e. water drawn out of the cistern and back down the rising main. Scalding could also result from a shower appliance connected direct to the rising main; showers therefore must always be fed from a cold-water storage cistern.

Many water authorities prefer the indirect system as:

1. It provides a large reserve in the event of mains supply failure.
2. It limits demand on the main, as the cistern can accommodate peak period requirements and refill slowly from the main therefore economising in mains pipe size and possible pumping costs.
3. Low pressure reduces noise within the system and wear on taps and other fittings.

Cisterns in the direct and indirect systems are of minimum actual capacities of 115 and 230 litres respectively. This refers to the quantities of water contained inside the vessel under the control of a

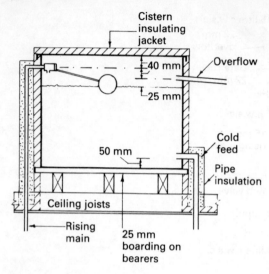

Fig. 6.4 Installation of a cold-water storage cistern

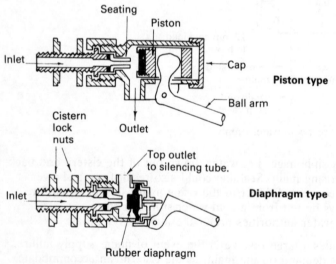

Fig. 6.5 BS ball valves

ball valve. Correct cistern installation as shown in Figs 6.4 and 6.5 illustrates the regulating mechanism of the British Standard pattern and the diaphragm ball valves.

Water control

Control of water distribution is either by stop valve or gate valve.

A stop valve is basically the same as a bib tap with a renewable washer to absorb wear. Rotation of the head slowly closes or opens supply, reducing possible damage by impact or water hammer. The gate valve operates similarly, but has a metallic gate or sluice instead of an absorbent washer. This will wear with use, and under high pressure could vibrate or fail to seat properly. It is much cheaper than a stop valve, and must only be used on low-pressure supplies, and never on the main. Drain valves are necessary at low points in the system in order to remove water when the building is unoccupied and to permit repair and maintenance work. Sectional details showing the function of these valves are shown in Fig. 6.6

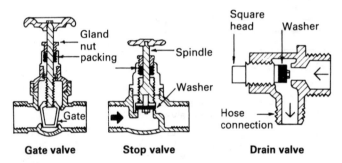

Fig. 6.6 Valves

Water supply to high-rise buildings

Multi-storey buildings frequently provide difficulties for cold-water supply beyond the lower floors. The local water authority will provide information on the minimum pressures expected in particular areas. From this data the maximum height achieved by mains pressure can be calculated. During peak demand daytime periods, the maximum mains pressure in most areas is likely to be between 150 and 300 kN/m^2 (kPa). Assuming the minimum figure, the maximum head supplied from mains pressure will be $\frac{150}{9.81} = 15.29$ m.

If floor-to-floor heights are 2.5 m, then $\frac{15.29}{2.5} = 6.1$ storeys, i.e. six storeys, can theoretically receive direct mains supply. However, designers prefer to leave a residual head of 5 m above the highest outlet, therefore for practical considerations the calculation would be based on a 10.29 m maximum head.

Drinking water and storage supplies required in excess of mains supply pressure must be boosted by pumps. Pumping direct from the main is unsatisfactory as the variable demand would impose considerably on pump efficiency, causing excessive wear. Additionally, if equipment fails there is the possibility of water being

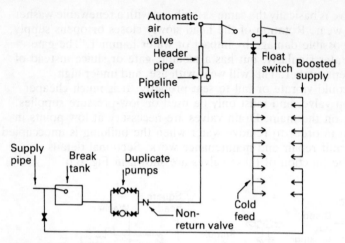

Fig. 6.7 Control by water level

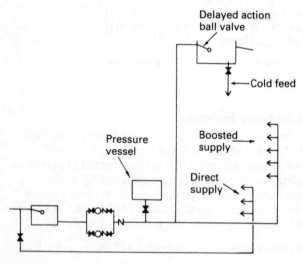

Fig. 6.8 Control by water pressure

drawn back into the main, hence most water authorities' insistence on the use of a break tank to provide positive separation. The two most acceptable systems of boosted supply are control by water level or by water pressure. The installation principles are shown in Figs 6.7 and 6.8.

Control by water level

This system has duplicate certrifugal pumps fitted after the break tank. One is operational and the other on standby. The pump is activated by level switches in the pipeline and the storage cistern. When a high-level drinking tap is opened, water is supplied from a header pipe or tank until the level drops to the pipeline switch. This introduces a pump to maintain supply at the taps and to refill the header until a time-delay switch disconnects the pump. A float switch in the storage cistern will also start the pump to replenish supply when the cistern is about half full.

In buildings over fifteen storeys, zoning of water supply is often preferred, to reduce the structural load of roof-stored water and to reduce the pressure in distribution pipes at lower levels. This is shown in Fig. 6.9 where drinking-water tanks provide a maximum head of 30 m.

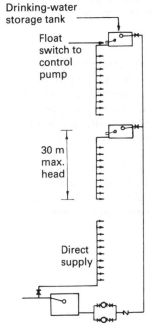

Fig. 6.9 Zoned supply to very tall buildings

Control by water pressure
This system is often preferred as it eliminates the need for electrical controls and wiring throughout the full height of the building. The system is permanently pressurised, therefore a header pipe is unnecessary, but a delayed action ball valve should be used in the storage cistern to regulate pumping. This device is shown in Fig. 6.10 with functional explanation. The pneumatic pressure vessel, compressor, electrical controls and pumps can be purchased as a

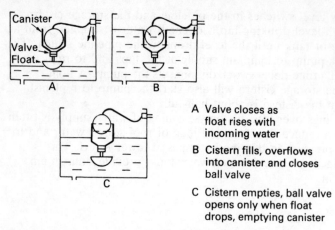

A Valve closes as float rises with incoming water

B Cistern fills, overflows into canister and closes ball valve

C Cistern empties, ball valve opens only when float drops, emptying canister

Fig. 6.10 Function of the delayed action ball valve

packaged unit thereby simplifying installation considerably. Adequate water pressure at taps and storage cisterns is maintained by a cushion of air compressed inside the pressure vessel. As demand reduces the pressure, a pressure switch introduces the pump to continue supply and to increase the air pressure contained in the vessel until the pressure switch stops the pump. Eventually some of the air cushion is absorbed by the water. To prevent excessive pump action, a float switch starts the air compressor to replenish the air volume. Figure 6.11 shows the operation of the pressure vessel, which, with the aid of delayed action ball valves, will considerably reduce pump wear, economising in maintenance and running costs.

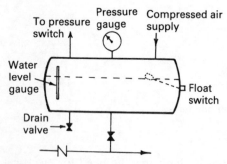

Fig. 6.11 Pressure vessel

Hot-water supply

Domestic installations

Hot water may be provided by centralised storage or by localised water heaters. Choice of method will depend on the size of dwelling and number of hot-water outlets, availability of fuel, space for equipment, capital expenditure, running costs and anticipated maintenance.

Centralised systems

Centralised systems have three major components, boiler, hot-water storage cylinder and cold-water storage cistern. The boiler, which may be gas, oil or solid fuelled, is normally situated at ground floor in close proximity to the storage cylinder at first floor. This is to reduce heat loss from the intermediate pipework and to provide efficient gravity circulation. The cold-water storage cistern provides the cylinder and boiler with water in the direct system, and a much smaller (36 litres) expansion cistern will also be required with the indirect system to supply the boiler and primary pipework only. The differences between direct and indirect systems are shown in Figs 6.12 and 6.13.

Direct system. A direct system may only be used where central heating is not required and in soft-water areas. Hard water contains calcium and is normally extracted from bore holes through chalk or limestone subsoils. If this water is used in a direct system of heating the boiler and primary pipework will 'fur-up' with calcium deposits, eventually causing blockage and boiler explosion. The extent of calcium precipitation will depend on the amount in the water and the boiler operating temperature. It is a relatively cheap installation and provides a more efficient heat transfer between boiler and cylinder than the indirect system.

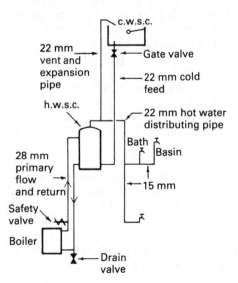

Fig. 6.12 Direct hot-water system

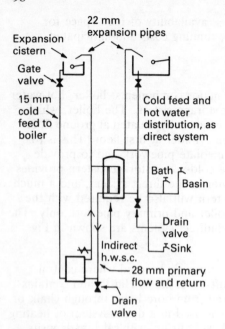

Fig. 6.13 Indirect hot-water system

Indirect system. This system is designed to provide the boiler with an independent water supply from a small expansion cistern. Water heated in the boiler circulates via the primary pipework to a coil or heat exchanger within the cylinder. This effectively seals the water in the primary circulation area preventing fresh water containing calcium from entering the boiler and associated pipework. Hot water for baths, basins, etc. is indirectly heated from the cylinder coil and replenished directly from the cold-water storage cistern. The amount of calcium precipitated here is negligible as water temperature is much lower than in the boiler. However, if an immersion heater is used in the cylinder when the boiler is not required, it can be expected to develop superficial lime scale and become less efficient.

Secondary circulation

Domestic hot-water installations rarely require circulation pipework between outlet and cylinder. It is sufficient to have short 'dead legs' of cold water from the storage cylinder to taps. The water authorities are less sympathetic with larger installations, and to restrict the quantity of cold water wasted before hot water is delivered, long hot-water supply pipes are not permitted unless secondary flow and return pipes are fitted (see Table 6.1).

Table 6.1

Pipe size (internal diameter)	Max. length (m) of hot-water supply pipe without a secondary return
Not exceeding 19 mm	
Exceeding 19 mm but not exceeding 25 mm	7.6
Exceeding 25 mm	3

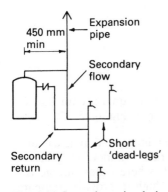

Fig. 6.14 Secondary circulation

Localised water heaters

Localised water heaters may be instantaneous or of limited storage. Instantaneous heaters have the advantage of no storage heat losses, but have limited flow facilities for large appliances such as baths. These units require a rapid transfer of heat energy, therefore gas and electricity are the only feasible fuel sources. Control of heat by thermostat and adequate insulation are provided by unit manufacturers to conserve fuel.

Electric water heaters

Electric water heaters may be categorised according to their method of heating water and mode of installation.

Pressure heaters. These are supplied with cold water from a storage cistern and may be wall mounted or conveniently situated under the sink (originally termed UDB's, i.e. 'under the draining board'). Capacities range from 50 to 450 litres, and the larger units are provided with two immersion heaters top and bottom, to provide

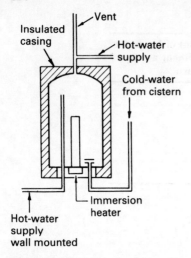

Fig. 6.15 Electric pressure type water heater

sufficient hot water for general use and a large quantity for baths. Installation and a sectional detail are shown in Fig. 6.15.

Cistern or cylinder/tank heaters. These are a units combining a cold-water storage cistern and a hot-water storage cylinder. They are most suited to flat roof dwellings and flats where water storage is difficult. They must be situated relatively high to provide adequate pressure to outlets on the same floor. Sufficient head for a shower is not likely to be possible. Installation may be from mains or cistern and is shown in Fig. 6.16.

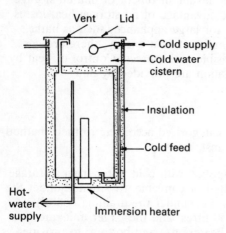

Fig. 6.16 Cistern type, electric thermal storage heater

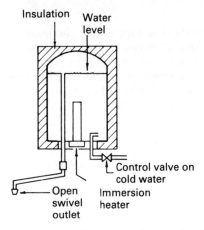

Fig. 6.17 Wall-mounted free outlet electric heater

Free outlet heater. These are most popular over a sink or basin, and are available in capacities up to 136 litres for use over a bath. They are either cistern- or mains-fed, without the need for expansion or overflow pipes. A sectional detail of a typical sink heater is shown in Fig. 6.17.

Instantaneous heaters. These are gaining popularity as shower fittings and for spray hand washing. A 3 kW heater is sufficient for basin use, providing a flow rate of about 1.5 litres/minute. A shower unit discharging about 3 litres/minute will require a heater rating of 6 kW. The principle is shown in Fig. 6.18 with a pressure switch designed to protect the element, preventing it switching on before the water flows.

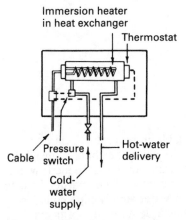

Fig. 6.18 Electric instantaneous water heater

Gas water heaters

Circulators. These are small boilers used to supplement the main heating source, or for use as an economy measure during summer months to provide stored hot water when the main boiler is not operational. Circulators occupy a very small space and may be fitted at the same floor level as the cylinder as shown in Fig. 6.19.

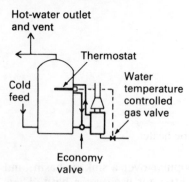

Fig. 6.19 Gas circulator

Storage heaters. These are commonly known as "multi-points", as they store sufficient hot water to supply a bath, basin and sink. They are also available with a swivel spout outlet in capacities up to 25 litres for use over a sink. Installation is shown in Fig. 6.20.

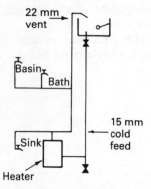

Fig. 6.20 Installation of gas storage heater

Instantaneous heaters. These are available in several capacities to suit sink, basin or bath. The larger units are rather bulky, unsightly and noisy in operation. When the hot water outlet is opened, water flow operates a gas valve, introducing burners which transfer heat to a cumbustion chamber having a finned water-supply pipe coiled around its surface. Figure 6.21 provides a detail of the principal components.

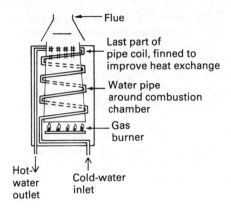

Fig. 6.21 Instantaneous gas water heater

Low pressure hot-water supply to high-rise and large buildings

The following systems illustrate the application of simple low pressure hot-water systems to moderately large premises. Figure 6.22 shows the extension of a normal indirect hot-water supply system to provide requirements for a hotel, hostel or similar residential building. It is shown on three storeys with a hot-water storage tank close to the taps and below the supply cistern level. The storage tank is particularly useful for satisfying large quantities of hot water

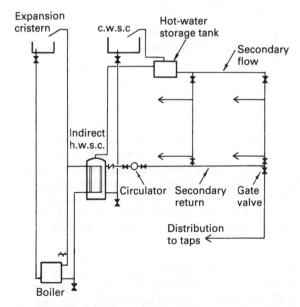

Fig. 6.22 Hot-water supply to a three-storey hotel/hostel

during peak demand periods. If the nature of the building is less demanding, the tank may be omitted provided the storage cylinder has sufficient capacity to meet requirements.

Precise calculation of hot-water storage requirements is determined from British Standard, Code of Practice 342. This document provides data relating to the building purpose and the type and capacity of sanitary appliances. Where a hot-water storage tank is not used, and storage requirements are high, parallel duplication of cylinders is possible. The installation is shown in Fig. 6.23 with the provision for duplicate boilers also. Repair and maintenance can be undertaken with minimum disruption as at least one cylinder can usually be left operational.

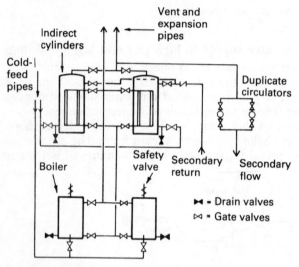

Fig. 6.23 Duplication of plant

Supply to high-rise buildings will also require duplication of plant, with storage cylinders linked to the boiler in series and cold-water storage cisterns arranged to provide a 30 m maximum head. This is shown in Fig. 6.24 with the provision of an automatic air release valve to reduce pipework and eliminate the expansion cistern. If the cistern is omitted an expansion vessel is situated on the primary pipework to absorb expansion of water from heating.

Solar heating of water
The principle of heat collection and storage transfer is shown in Fig. 6.25. Many systems have now been installed in the UK, but their effectiveness and usefulness is limited by uncertain climatic

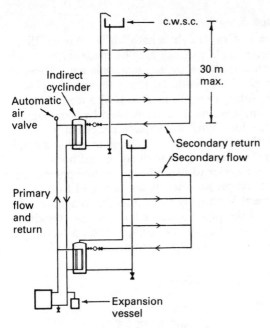

Fig. 6.24 Centralised hot-water supply to high-rise buildings

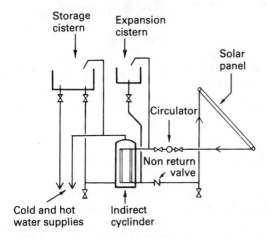

Fig. 6.25 Solar-heating principles

conditions. Initial installation costs are recovered by saving in the use of conventional fuels, but the payback period is likely to be very long. Optimum collector area for domestic use is between 4 and 6 m², and the south-facing roof should ideally pitch at about 40°.

Jointing of copper tube

Copper tube is available in 6 m straight lengths or in coils to BS 2871 : Part 1. Joints, bends and branches may be made with soldered fittings, either manipulative or non-manipulatives. These are illustrated in section in Fig. 6.26. Soldered joints require the pipe surface and inside of the fitting to be cleaned with wire wool. Flux is applied to the jointing surfaces and heat introduced by blow lamp. Soft solder is fed into the joint until it completely spreads by capillary action. This is the cheapest type of coupling and some are provided with an integral solder ring. Manipulative and non-manipulative couplings are easier to use, but more cumbersome and expensive. Manipulative compression joints are specified for use below ground as movement is unlikely to disturb the swaged bead formed around the end of the pipe, contained within the coupling. Non-manipulative joints have a brass ring compressed on to the pipe surface by a hexagonal nut to secure the joint against leakage.

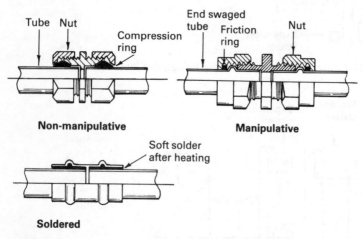

Fig. 6.26 Copper pipe couplings

Gas and electricity supply

Gas

Gas mains form a network of service pipes generally situated below the roads. Steel pipes wrapped in a protective grease tape have been used, prior to the recent introduction of distinctive yellow uPVC pipes which are proving more convenient and easy to use. uPVC pipes cannot be used for internal distribution of gas as the exposed pipework has inadequate fire resistance. Internal pipework is either mild steel to BS 1387 : 1967 or copper to BS 2871 : Part 1 : 1971.

Under the provisions of the 1972 Gas Act, owners of premises situated within 25 yards (23 m) of a gas main have a statutory right to gas supply if they require it.

Co-ordination of service pipe location inside the boundary and position of gas meter and controls is agreed by consultation with the regional gas authority. A typical domestic installation is shown in Fig. 6.27, where a 25 mm bore service pipe is quite sufficient with a supply pressure of between 1 and 2 kN/m^2. When several dwellings exist within one building, and more than one meter is required the service pipe diameter will be increased accordingly (see Table 6.2).

Table 6.2

Nominal bore of pipe (mm)	No. of dwellings/meters
32	2–3
38	4–6
50	Over 6

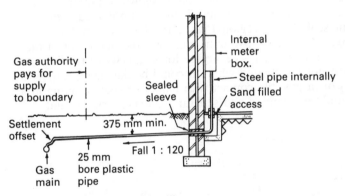

Fig. 6.27 Gas supply to a dwelling

Variation exists to accommodate the number and nature of appliances and size of dwelling. Accurate determination of pipe size is necessary for industrial installations and in other situations where numerous appliances are fitted. Design tables are provided in BS, CP 331 : Part 3 : 1974 for this purpose.

Installation to houses

The service entry to a house is similar to the water main entry, except that isolating valves are not necessary and the ground cover is reduced, as the moisture content of natural gas is negligible and therefore unaffected by frost. A slight slope back to the main is provided on the service pipe in order to dispose of any water that

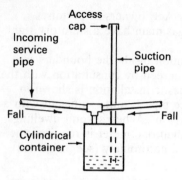

Fig. 6.28 Condensate receiver

may enter the pipe. If this is impractical, a water trap or condensate receiver of the type shown in Fig. 6.28 should be installed.

Service pipe entry to the building must be protected against settlement and never taken below the foundation. Ideally, the service-pipe should enter on the same side as the main and if possible avoid penetration of the wall as shown in Fig. 6.29. A protective plastic sleeve of 75 mm diameter is provided with each end mastic sealed. The regional gas authorities now prefer plastic meter boxes, 620 mm × 540 mm, built into the outer leaf of the cavity wall with external accessibility for simplicity of reading. A meter must not be situated where it could be damaged, subject to dampness or in possible contact with flames or sparks. Siting in a garage should be avoided and separate chambers for gas and

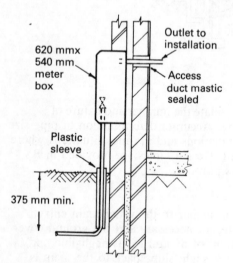

Fig. 6.29 External meter box

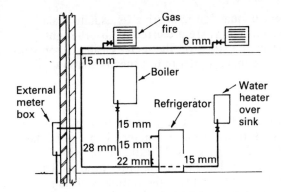

Fig. 6.30 Internal gas distribution to a house

electricity meters will be necessary. Siting under a stairway is possible, if the surrounding construction has at least half-an-hour fire resistance and a self-closing door.

Interior pipework could follow the typical installation shown in Fig. 6.30. All the pipework must be adequately supported with purpose-made brackets and protected with incombustible material where passing through a wall or floor (see Fig. 1.10, p. 10). Pipe runs in the wall cavity must be avoided and electrical cable should not be supported or taped to a gas pipe. Pipes must also avoid any source of heat.

Meters are the property of the gas board. They are fitted with either conventional inlet and outlet connections or a more compact annular connection shown in Fig. 6.31. The control valve operates with a simple 90 ° turn to isolate or open supply. The pressure governor is adjusted by a gas board official to suit the appliances

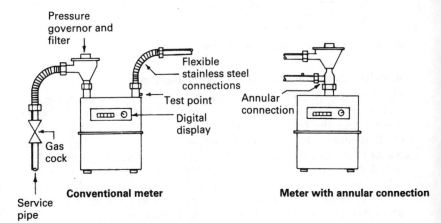

Fig. 6.31 Domestic gas meters

installed and, to prevent unqualified tampering, is usually sealed. Flexible stainless steel connections are preferred to absorb slight movement and prevent undue stress on the meter casing. A pressure test point is situated on the meter outlet to test the installation pipework for leakage.

Testing

To test, all appliances and valves must be shut off before air is pumped into the pipework through the test point by means of the apparatus shown in Fig. 6.31. The glass 'U' tube or manometer is filled with water to the zero level and pumping introduced until a pressure of 300 mm water gauge or twice the normal working pressure (whichever is greater) is recorded. Ten minutes is allowed for adjustment of air temperature, before the pressure is re-applied and observed for a further 15 minutes. Any fall indicates leakage and this may be located by application of soap solution over the joints until the defect is noticed by appearance of bubbles.

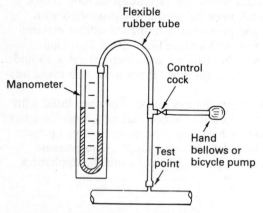

Fig. 6.32 Installation testing equipment

Installation to flats and other large buildings

Supplies to industrial premises and all other supplies requiring a service pipe over 50 mm bore are provided with an accessible service valve where the pipe enters the property boundary. This must be easily identified for closure in event of fire. Some installations will require several meters, particularly blocks of flats and hostels. The installation principles are shown in Fig. 6.33. If the building is of sole ownership and divided into flats for letting, the landlord will probably require secondary pre-payment meters to each flat and a primary meter to record and check total consumption.

Industrial meters have flanged connections to suit 75, 100, 150 or

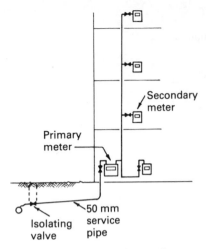

Fig. 6.33 Gas installation to flats

200 mm bore pipes. Connections and accessories are similar to domestic meters except flexible couplings are unnecessary. With gas board approval, a by-pass is permitted to provide supply continuity if the meter becomes defective

Flues for gas appliances

The function of a flue is to permit the safe extraction of burnt gas and to provide sufficient draught to encourage efficient fuel combustion. Natural gas combustion introduces a very low sulphur dioxide content to the atmosphere, easily satisfying the requirements of The Clean Air Act. Possible exceptions would only be exceptionally large gas-fired installations.

Domestic installations

Flue requirements for gas fires and boilers of input rating less than 45 kW (Building Regulations, Part L – Class II appliance) are much simpler to provide than for oil or solid fuel appliances. The intensity of heat is less and the combustion products less toxic.
Gas combustion does not produce smoke and as water vapour will only be observed as steam during cold weather it is not necessary to provide a tall chimney. Hence the convenience of a low level balanced flue shown in principle in Fig. 6.34. Appliances connected to these flues are 'room-sealed', as both the air for combustion and from the exhaust are external. As the flue inlet and outlet are so close, air pressure is maintained at a constant rate whatever the wind conditions. Where appliances are situated on an internal wall, a conventional flue or flue block system will be necessary.

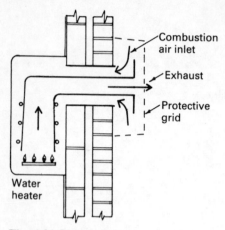

Fig. 6.34 Balanced flue

Conventional flue

A conventional flue is considerably more expensive to install than a balanced flue. Part L15 and L16 to the Building Regulations defines acceptable flue pipe materials:

(a) Socketed clay pipes to BS 65 and 540 jointed and pointed with high alumina cement mortar.
(b) Socketed cast iron to BS 41, with an internal coating of vitreous enamel.
(c) Sheet metal to BS 715.
(d) Stainless steel.

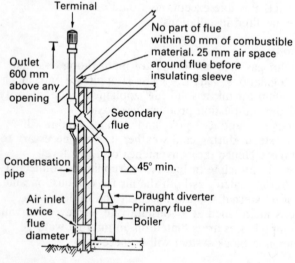

Fig. 6.35 Conventional flue

(e) Asbestos cement to BS 835 or BS 567, coated internally with an acid-resisting compound.

All spigot and socket connections must be made with the socket uppermost. A typical installation is shown in Fig. 6.35. A draught diverter is used primarily to prevent down draughts interfering with combustion and to reduce excessive flue pull. This is shown in Fig. 6.36, with the diluting effect of secondary air reducing both the CO_2 content and flue gas temperature.

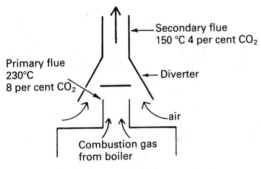

Fig. 6.36 Effect of draft diverter

Pre-cast concrete flue blocks for gas fires

A standard section flue block is illustrated in Fig. 6.37 with installation principles. Offsets are also available for flue diversion. Building Regulation L14 (1)(b) specifies that these blocks must be made of high-alumina-cement dense concrete and jointed with hac mortar. They are most suitable for use with gas fires, and must be installed in accordance with these additional Building Regulation requirements (see Table 6.2).

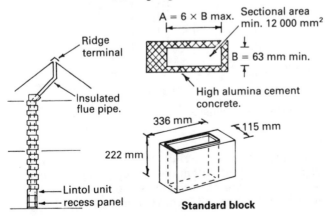

Fig. 6.37 Gas fire flue blocks

114

Table 6.3 Building Regulation L14

Size of flue	Maximum length
A, less than 3 × B	21 m on internal wall 11 m on external wall
A, greater than 3 × B	12 m on internal wall 6 m on external wall

Shared flues

Shared flues are provided to simplify the disposal of burnt gas from numerous appliances in multi-storey buildings. Building Regulation M10 provides details relating to the constructional size of flue and additional safety requirements. These are mainly the fitting of draught diverters, flame failure devices to prevent unburnt gases entering the flue and, where room-sealed units are used, the provision of air at the uppermost appliance containing not more than 2 per cent CO_2. Three possible systems are shown in Fig. 6.38.

The Se-duct (developed by the South East Gas Board) and the U-duct are ideal for room-sealed balanced flue appliances. The former provides for location of appliances either side of the duct but requires a combustion air inlet at ground level. The U-duct has a high-level air inlet, which is useful where ground access is difficult but will occupy greater space within the building. The shunt duct is the most economical means of providing a combined flue for conventional flued appliances.

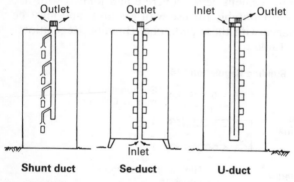

Fig. 6.38 Flues to multi-storey buildings

Fan-diluted flues

Fan-diluted flues are particularly useful where it is desirable to contain the length of flue pipe within one storey. They are very often employed in shops with offices or flats above, where it would be difficult and unsightly to install a conventional natural draught

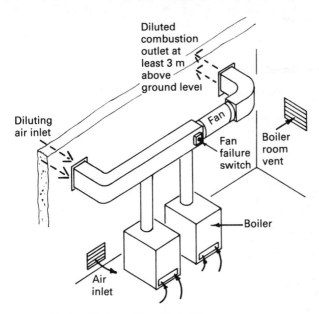

Fig. 6.39 Installation of fan-diluted flue

flue. To allow a balanced effect the inlet and discharge louvres should be on the same wall with the outlet at least 3 m above ground level to provide clear headroom. Because of this relatively low level of discharge the effect of fresh air dilution must be to reduce the CO_2 content to below 1 per cent. Figure 6.39 shows a diluted-flue installation, with a fan failure device to protect the boiler.

Electrical installations

Electricity is generated by steam turbines powered by oil, coal or nuclear fuels. It is distributed via the national 'super' grid at 400 kV, transforming to networks of 275 kV, 132 kV, and 33 kV for underground urban allocation. Local transformers and sub-stations make a further reduction to 11 kV for large industrial sites and 415 V for general distribution. The 415 V, 50 cycles/second, 3-phase supply contains four conductors, three phase wires and a neutral which is earthed at the sub-station. Connection between any two-phase wires provides a 415 V supply, suitable for heavy machinery. Connection between a phase wire and the neutral produces a potential of 240 V. Buildings receiving a 415 v supply often have all three phase wires for the convenience of distribution, hence the term, 3-phase supply. Dwellings having only one-phase wire receive single-phase supply. Figure 6.40 is a simplification of the three-phase and single-phase distribution principle.

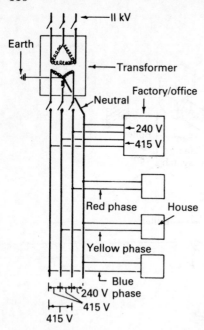

Fig. 6.40 Three-phase and single-phase supply

Supply to small buildings

An underground electricity supply cable containing one live phase wire and a neutral conductor is provided by the electricity board. This is left close to the outer wall with sufficient cable for connection to a meter. Existing supplies are often overhead but the majority of new housing will be provided with a supply buried in a trench, 450 mm below the surface. The intake is arranged with access through the wall cavity or other suitable duct as detailed in Fig. 6.41. The cable is PVC insulated and copper or aluminium sheathed to provide continuity of earth from the meter and consumer unit. If the sheathing cannot provide sufficient earth or if the supply is overhead, a separate earth electrode is driven into the ground and provided with a cross-bonded connection to the gas and water service pipes (unless plastic). This is shown in Fig. 6.42, the purpose being to reduce the effect of electric shock from faulty equipment such as an immersion heater or heating pump. Furthermore, all metal fittings, e.g. bath or sink with the absence of metal-to-metal joints must be similarly bonded.

Protective multiple earthing (PME). Metal sheathing generally provides an excellent earth, but its efficiency can vary due to differing ground conditions. Because of these earth resistance fluctuations PME is becoming more widely used. Figure 6.43

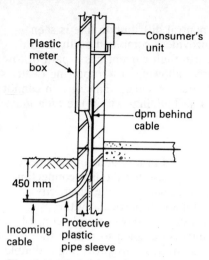

Fig. 6.41 Electricity supply to a meter box

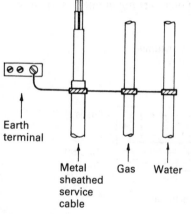

Note: Bonded connection to gas pipe to be within 150 mm of outlet flexible connection.

Fig. 6.42 Cross bonding of services

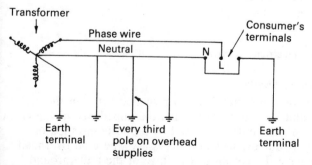

Fig. 6.43 Installation of PME

illustrates the concept where the neutral distribution line is seen connected to earth electrodes to provide a combined earth/neutral conductor. This ensures that an earth fault current has a path of low resistance back to the supply source, allowing easy rupturing of fuses or operation of an excess current circuit breaker. This system cannot be used on individual dwellings as all buildings within the area must conform to this system.

Distribution

The incoming service cable is terminated in a sealed compartment containing a neutral link and a 60-amp fuse for the live connection. This fuse is sealed, and may only be replaced by the electricity authority. It protects the entire installation, particularly the consumer's unit and meter. The consumer's power supply control unit (abbreviated to consumer's unit) is designed to reduce the hazard of individual circuit switches and fuses by providing a double pole switch for complete isolation from the mains supply, neutral and earth terminal blocks and live fuse ways or miniature circuit breakers for individual circuits. A typical arrangement is shown in Fig. 6.44.

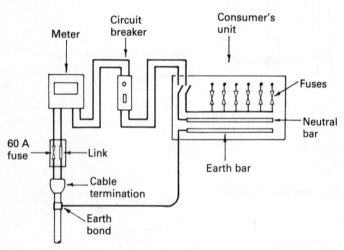

Fig. 6.44 Domestic electricity controls

Earth leakage circuit breakers

A circuit breaker or trip switch is often provided between the meter and consumer's unit. It is necessary when the impedance of the earth fault is too high, which could be caused by a poorly installed or ineffective earth conductor. It is also necessary where an overhead supply does not have the earthing advantage of the underground metal-sheathed cable.

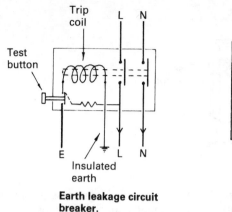
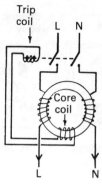

Earth leakage circuit breaker.

Current operated circuit breaker

Fig. 6.45 Circuit breakers

Voltage-operated ELCB. The operating principle is shown in Fig. 6.45. The trip is energised by an electro-magnetic force when the potential at earthed metalwork reaches 40 volts to create sufficient flow of current in the coil.

Current-operated ELCB. This is also shown in Fig. 6.45. The live and neutral conductors are coiled on a circular frame. During normal operation the currents flowing in the live and neutral are the same, therefore the magnetic fluxes are balanced. If an earth fault occurs, more current flows in one conductor, causing an imbalance which generates an electromotive force in the core coil to operate the trip. This is also limited to a 40 V maximum potential at earthed metalwork.

Fuses

To prevent fire and electric shock it is necessary to provide a means of interrupting the flow of excessive current. The simplest and cheapest method is provision of a rewirable fuse at the live supply point of each circuit.

This is not the most efficient form of protection, as there is always a time delay before the metallic element ruptures. Cartridge fuses are better, but more costly. These have a ceramic barrel or tube with metal end caps and a silver wire stretched between them. When an overload occurs, quartz filling around the wire vaporises the centre section leaving a solid non-conducting core.

The relationship between the rating of a fuse and its rupturing capacity is known as the fusing factor:

$$\frac{\text{fusing current}}{\text{current rating}} = \text{fusing factor}$$

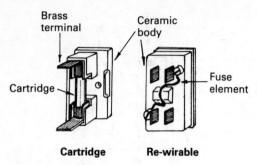

Fig. 6.46 Fuses

Re-wirable fuses have a fusing factor of up to 2. Vagueness is caused by various considerations, either the element material, its cross-sectional area, length and whether it is open to air or insulated. Cartridge fuses have a fusing factor of between 1.2 and 1.6. Variation with each fuse type also occurs because its purpose is to destroy itself, therefore it is impossible to test individual fuses. A circuit breaker however performs the same function as a fuse as an automatic switch with indefinite use. Most important, it can be tested after manufacture to ensure a fusing factor close to unity. Their use in miniature form as a replacement for fuses has been limited by production cost, but they are gradually becoming more widely employed. They function on similar principles as the ELCB, with an electromagnetic force rapidly opening the contacts. Thermal trip circuit breakers have also been produced, having a bi-metallic strip which expands as the temperature increases with current overload, to operate a contact-breaking mechanism.

Circuits and cables

Electricity is conducted through copper- or aluminium-cored cable surrounded with a colour-coded PVC insulation and a grey or white outer sheath. For most simple installations the cable contains a live (red insulation), a neutral (black insulation) and an uninsulated earth as shown in Fig. 6.47. When the earth is exposed it must be sheathed with green PVC for easy identification and protection against unnecessarily earthing a circuit. The size of conductor may be determined by calculating the maximum current it will be expected to carry. For simple installations used in dwellings, the cable ratings shown in Table 6.4 are suitable:

Exposed wiring clipped as shown in Fig. 6.47 is the easiest and quickest method of installation. It is undesirable where appearance is important and, from the safety aspect, unacceptable in most industrial situations. Concealed cables should be protected by a cover strip which permits sufficient space for the cable to be removed if re-wiring or modification is necessary. Steel or plastic

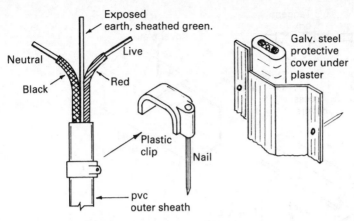

Fig. 6.47 Electric cable, methods of securing

Table 6.4

	Diameter (mm)	Sectional area (mm²)	Fuse rating (Amps)	Limitations
Lighting	1.13	1	5	Max. 10 light fittings
Power sockets	1.78	2.5	30	Max. 100m², not in bathroom
Immersion heater	1.38	1.5	15	Butyl-coated flex from fused socket to heater
Cooker	1.04 × 7	6	30 or 45	Fuse rating depends on cooker rating.

tube conduits are preferred for industrial and commercial buildings, steel having the advantage of continuous earth provision. Conduit boxes are shown in Fig. 6.48; these provide access for cables and modifications that may be required at a later date. Trunking is also popular for industrial installations as an alternative to a very large conduit or several small conduits. Compartmentation provides easy segregation and identification of service cables. An example of steel trunking and a variation, available in plastic which doubles as skirting, are shown in Fig. 6.48.

Lighting
Light fittings are usually installed with a 'loop-in' system shown in

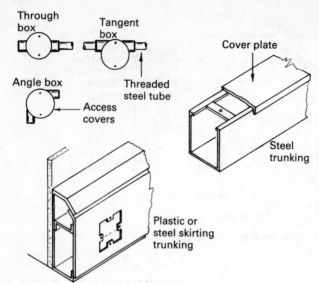

Fig. 6.48 Conduit and trunking

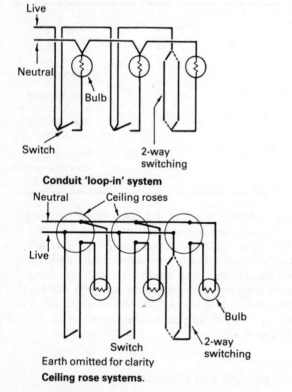

Fig. 6.49 Wiring to light fittings

Fig. 6.49. In housing, separate circuits are normally provided at each floor level, each protected by a 5-amp fuse at the consumer's unit, and each circuit limited to ten fittings. The following theoretical calculation indicates that if all ten fittings of 100 W rating were installed and in use, the fuse rating would have a very narrow safety margin:

$$\frac{\text{Watts}}{\text{Volts}} = \text{amps} \qquad \text{hence,} \qquad \frac{10 \times 100}{240} = \underline{4.16 \text{ A}}$$

Figure 6.54 also illustrates the same installation using conduit with single-cored cable. Both lighting details show two-way switching for use in hall/landing and bedroom situations.

Power circuits
Socket outlets should be placed so that only a short flex is needed between socket and appliance. In kitchens and workshops they should be between 150 and 250 mm above the work surface, and for general use at least 150 mm above floor level. In buildings designed especially for the old and disabled, the socket height is recommended between 750 mm and 900 mm above floor level. Plugs are now standardised to BS 1363, having three rectangular section brass pins with a 13 A cartridge fuse carried on the live terminal. This plug is suitable for appliances rated up to 3 kW. For those rated below 720 W the 13 A (brown) fuse should be replaced by a 3 A (blue).

Sockets should be provided generously and preferably of the double pattern. The following details indicate the minimum number acceptable by local authorities for housing and improvement grant work:

Kitchen	4
Living area	3
Dining area	2
Bedrooms	2 each
Hall and landing	1
Garage	1 (optional)

One ring circuit of the type shown in Fig. 6.50 is limited to 100 m² of floor area, as resistance to the flow of electricity by long cable runs could cause a voltage drop below the maximum acceptable of 2½ per cent. With the exception of small dwellings, one ring main should serve each storey. Non-fused spurs may be added for convenience or when a structure is extended. The cable must be the same rating as used on the ring circuit, and connections must be made at a socket or with a 30 A junction box. The maximum number of spur sockets must not exceed the number of sockets and fixed appliances connected direct to the ring.

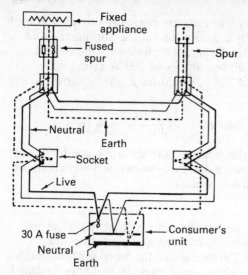

Fig. 6.50 Ring main to power sockets

Distribution in large buildings

Hospitals, blocks of flats, factories and offices have a higher electrical load than dwellings, therefore the 3-phase supply is taken direct into the building. Where the load is exceptionally high, a private sub-station is provided. Distribution to lighting and power

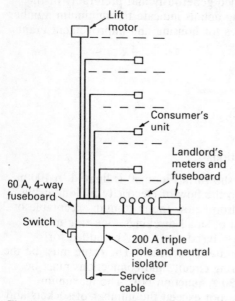

Fig. 6.51 Supply to a block of flats

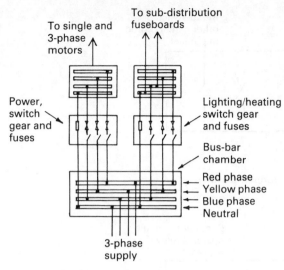

Fig. 6.52 Supply to a factory

fittings compares with smaller buildings, but is provided from numerous separate circuits with means of isolation and fusing at appropriate intervals. The diagrams in Figs 6.51 and 6.52 show radial distribution to a block of flats and a small factory. A large site will be more efficiently served with a 3-phase ring supply, commencing and returning to a private sub-station.

Installation work should be conducted under the Institution of Electrical Engineers Wiring Regulations, 15th edition, 1981. These are designed to promote safe and efficient installation of electrical equipment and services in all buildings.

Communication systems

Burglar alarms

Burglar alarms are usually independent of normal electricity supply systems, having battery sources of energy. Wiring systems contain two circuits with a common polar sensitive relay. These systems are either open or closed, as shown in Figs 6.53 and 6.54, respectively. Open systems have several contacts wired in parallel. When the contacts are connected the circuit is complete. This energises the relay which completes the alarm circuit to provide an audible signal. Closed systems operate conversely. Contacts are wired in series. When contacts are broken, the relay de-energises to complete the alarm circuit.

Detection devices are numerous, some highly scientific, such as beam optical systems and ultrasonic-sensitive instruments.

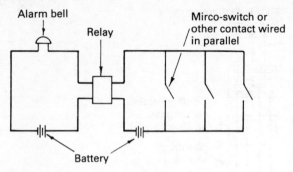

Fig. 6.53 Open circuit burglar alarm

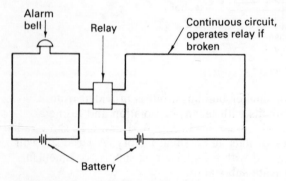

Fig. 6.54 Closed circuit burglar alarm

The simplest devices include micro-switches located between door and frame or window sash and frame, and pressure switches located under floor coverings. Another simple technique is location of a continuous fine copper wire conductor in walls, ceiling and floor. An intruder breaking through the structure fractures the wire which opens a closed circuit.

Telephone installation

Telephone services cannot be provided by tapping off a main, similar to gas, water and electricity. Each telephone requires two wires from the telephone exchange. British Telecom endeavour to anticipate future demand and local installations invariably contain ample spare cable.

Supplies to buildings are either underground or overhead, depending on existing location of cable. Figure 6.55 illustrates both possibilities with provision of earthing by metallic water service pipe or earth spike. Copper earth wire is at least 1.5 mm^2 and this may be concealed under plaster, in the wall cavity or within the floor structure.

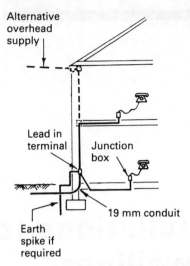

Fig. 6.55 Domestic telephone installation

Whether the supply is underground or overhead, the builder will be required to provide a 19 mm conduit for easy access. Underground supplies may be accommodated in a common services trench or a separate trench of at least 375 mm depth.

Chapter 7

Heating installations to domestic dwellings

There are three popular methods of providing thermal comfort to dwellings. These relate to the type and choice of heat emission; radiators, ducted warm air or a fabric heating element.

Fuel is also a very important consideration and the choice may be limited by availability. Gas, oil and solid fuels are the most common fuels used for central heating installations, although electricity is possible but very expensive when used as a direct energy source. The use of off-peak electricity is an economical consideration where other fuels are not convenient, but control and regulation of heat output to suit variable conditions and requirements is difficult. At present the development of renewable energy resources, such as wind, tides, sun and wave power, are not economically viable, but with the increase in conventional fuel costs optimism exists that further research into their potential may be harnessed as a useful fuel source.

Natural gas is currently more economical than oil or coal as a boiler fuel. Demand for gas supplies is increasing in advance of production and estimates suggest another 40 years' supply remains from existing sources. It is anticipated that additional sources will be located, considerably extending the supply period. If natural gas supplies are exhausted, there remains considerable coal deposits in the UK for returning to production of coal gas. Apart from relative cheapness, gas has the advantage of no storage requirements and suitability to automatic control.

Until the early 1970s when the price began to escalate, oil compared quite favourably with gas as a central heating fuel. Now its advantages are limited to use in remote situations where gas is not available, and its suitability to automatic control. Oil resources appear to be sufficient for about 100 years and current exploration is proving quite rewarding although the search is becoming more difficult, requiring advanced engineering techniques for its location and removal. The main disadvantage apart from cost is the necessity for storage facilities. Two grades are provided for central heating boilers, C2 or Kerosene used in vaporised form, and D, a gas oil similar to diesel oil for use with pressure jet injection. C2 is sometimes known as 28 second and D 35 second, both of which refer to the Redwood scale of viscosity.

The efficiency and cost effectiveness of solid fuel varies with the quality and type of material burnt. Log burners with a back boiler provide one of the cheapest means of heating water but require constant attention, and the heat exchange is unlikely to provide any more than supplementary energy to a main central heat source. The most efficient solid fuel for central heating boilers is anthracite, which has a higher calorific value than other solid fuels and low volatile smokeless emmission. The possibilities of automatic control are less likely than with the constant piped supplies of oil or gas. Some modern solid-fuelled boilers are equipped with a hopper feed system and sufficient storage for several days uninterrupted use. Ash deposits will require periodic removal, although the amount will be minimal if high-grade fuel is used. Coal reserves are known to be sufficient for several hundred years, but its economic use as a central heating fuel is dependent on its political value. The main disadvantages are provision of storage, handling and ash disposal.

Systems

Radiators

Traditionally, radiators were manufactured from cast iron in columns or panels. Modern radiators are pressed steel in single or double panel form. Variations exist with finned backing or in convector form to increase the heat exchange area. Some examples of radiators are shown in Fig. 7.1 with an illustration of the connections.

Heat is emitted from a radiator by convection and radiation. Up to 40 per cent is radiated, half of which is directed onto the rear mounting wall. To reduce this potential heat loss, particularly on an external wall, some form of reflective backing is necessary. A small sheet of foil-backed plasterboard mounted between the radiator brackets improves radiator efficiency considerably.

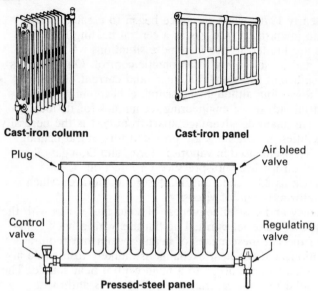

Fig. 7.1 Radiators

Low pressure hot-water heating systems

Domestic heating systems operate with a maximum boiler water temperature of about 80 °C. They also operate with an open vent pipe to prevent pressurisation and generation of steam. They may be classified as:

(a) Gravity systems, where circulation is due to convection currents caused by temperature and density changes.
(b) Accelerated systems, where circulation is provided by a pump.

Domestic gravity central heating systems are now obsolete except in small dwellings where circulation pipework is simple and the number of radiators minimal. The disadvantages compared with accelerated systems are many:

(a) careful positioning of pipe runs and radiators to encourage circulation;
(b) a 20 °C temperature difference between boiler flow and return pipes (80° – 60°), resulting in lower mean temperatures;
(c) large pipe sizes;
(d) limited thermostatic control.

The most significant difference between accelerated and gravity systems is the provision of a water circulator, commonly referred to as a pump. This provides several advantages:

(a) simplified pipe runs (relative levels are less important);
(b) smaller pipe sizes;

(c) a 10°C temperature difference between boiler flow and return;
(d) higher heat emission from radiators.

Both gravity and accelerated systems can be either one-or two-pipe systems. There are many variations of each and some are shown in Fig. 7.2 and 7.3. Two-pipe systems involve more pipework and are more expensive to install, but offer greater working efficiency, because one-pipe systems have three major disadvantages:

1. The control of one radiator noticeably affects the heat output of others.
2. Hot water passing through a radiator cools and returns to the

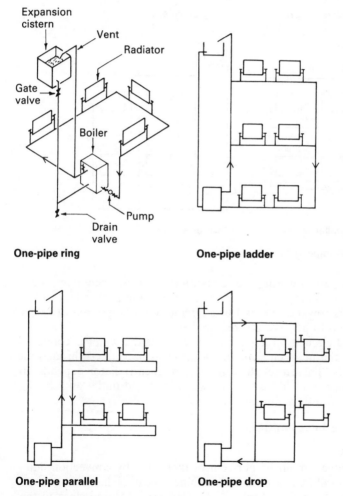

One-pipe ring

One-pipe ladder

One-pipe parallel

One-pipe drop

Fig. 7.2 One-pipe heating systems

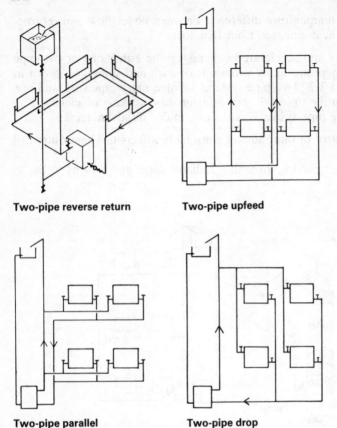

Two-pipe reverse return

Two-pipe upfeed

Two-pipe parallel

Two-pipe drop

Fig. 7.3 Two-pipe systems

main circuit, lowering the temperature of water supplying the following radiators. This distinctly affects the performance of radiators towards the end of a circuit and a large area of radiator is needed to achieve the required output.

3. Pumped circulation has limited success as the water is conveyed around the main circuit with ease, but is not forced through the radiators. Therefore, the efficiency of each radiator depends on the temperature difference between inlet and outlet and the convective effect this generates.

Alternative heat emitters

Convectors

These produce about 95 per cent of their heat by convection and comprise a cabinet or casing containing a heat exchanger element constructed from finned tubes. The element is located close to the

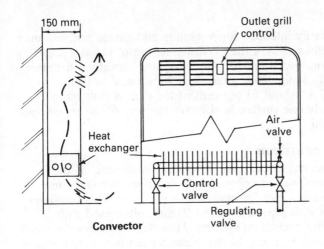

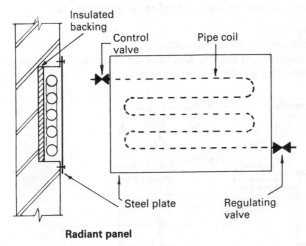

Fig. 7.4 Convector and radiant panel emitters

bottom of the casing to maximise the stack effect on the column of
warm air generated. The air flow will increase in proportion to the
height of the cabinet up to a maximum of about 1 m when resistance
to air flow will have an effect. An example of a wall-mounted
natural convector is shown in Fig. 7.4. These are also manufactured
as a skirting variation which provides well-distributed heat but
inefficient use of a finned tube element. Fan convectors are also
produced with a heat exchange element close to the top of the
cabinet. The fan draws cool air from the base of the cabinet and
projects it through the element, directing it horizontally across the
room. The advantages over natural convection are a higher heat
output and rapid response to heating a room.

134

Radiant panels

These are primarily intended for workshop and garage use, but may be used in dwellings where the obtrusive nature of radiators or convectors is considered undesirable. They are basically a flat steel plate attached to a pipe coil, fitted flush with a wall or ceiling as shown in Fig. 7.4. About 65 per cent of the heat is emitted by radiation; as only one surface is effective the rear of the recess must be well insulated.

Mini-micro-bore systems

Conventional heating circuits are known as 'small-bore' as the copper pipes employed are only 28 mm or 22 mm for flow, and return mains with 15 mm for branches to radiators. This compares very favourably with pipe sizes up to 50 mm, often used with traditional gravity circulation systems. However, 'mini-bore' systems using pipes of only 6, 8 and 10 mm diameter are now very common and offer many advantages over 'small-bore' installations:

(a) quick installation;
(b) minimal disruption to the structure;
(c) low water content, therefore fast heat-up rate, more efficient use of the boiler and better thermostatic response;

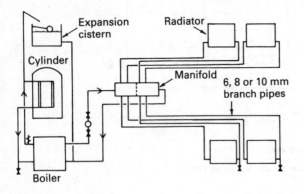

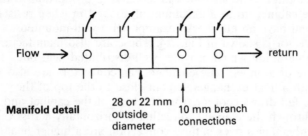

Fig. 7.5 Mini-bore system

(d) less conspicuous and neater appearance than other pipe systems;

(e) continuous lengths (soft copper pipe in coils), fewer joints and less chance of leakage;

(f) fast water circulation, therefore relative pipe levels unimportant as there is no chance of air locks.

Installation

Heating flow and return pipes are generally 22 mm, possibly 28 mm for large dwellings. These supply a manifold with several 'mini-bore' flow and return branches, or there may be separate flow and return manifolds. The manifold should be centrally located to provide fairly uniform distribution of branch lengths to radiators. Small terraced dwellings and bungalows will require only one manifold and larger buildings will be better served with one on each floor. Alternatively, a conventional 22 mm flow and return pipe may be provided with 10 and 8 mm branches at intermittent intervals to suit radiator positions.

Sealed systems

Sealed or closed heating and hot-water systems are gaining popularity in domestic dwellings as they offer advantages to the installer and user. The installation is cheaper than open vent systems as there is no expansion pipe or expansion cistern and cold feed to provide – only an expansion vessel. Also the boiler position is not critical; it can be located in the roof as the system is independent of gravity for circulation and filling. The expansion vessel has a nitrogen or air cushion separated from the water by a rubber

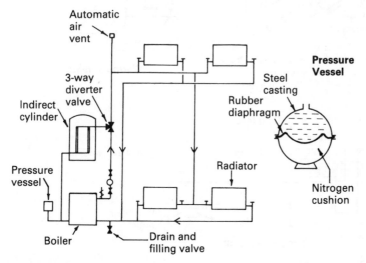

Fig. 7.6 Sealed heating system

diaphragm, which pressurises the water to permit a higher water temperature without conversion to steam. This permits smaller pipes and heat emitters, but radiators must be specified for use under pressure. The drain plug also functions as the filling point from a mains tap and hose pipe. The installation principles are shown in Fig. 7.6.

Ducted warm air

A ducted warm-air system is principally the same as a fan convector with distribution ducts attached to the warm-air outlet. The central heat source either contains a sealed combustion unit and flue as shown in Fig. 7.7 or a heat exchange battery provided with hot water from a separate boiler source, also shown in Fig. 7.7. Air is directed over the heater unit and pressurised through the delivery ducts by a centrifugal fan. This also creates a negative air inlet pressure to receive the cooler room air for re-heating. Remote areas

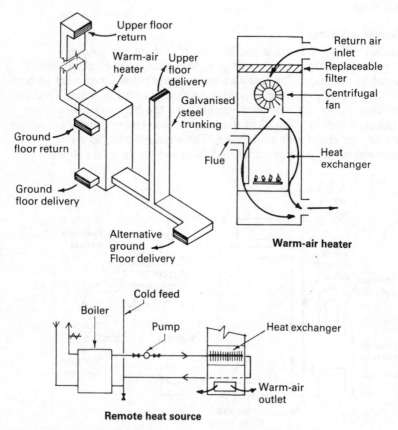

Warm-air heater

Remote heat source

Fig. 7.7 Warm-air heating

may have return air ducts to regulate distribution. Ducting is in circular or rectangular galvanised steel terminated with diffuser grills at floor level. Diffusers are normally fitted with a damper to regulate air flow and balance distribution. Alternatively, a damper is located in an accessible part of the ducting.

A variation, known as Electricaire, uses mainly off-peak night rate electricity to heat insulation blocks in a thermal storage unit. A core thermostat prevents overheating and a time switch controls fan distribution of ducted warm air.

Fabric heating

Fabric heating refers to the use of embedded heating elements in the floor, wall or ceiling structure. It is often known as invisible heating as there are no obvious heat emission surfaces, which permits greater flexibility in room presentation and planning. The elements may be polypropylene, steel or copper pipes conveying hot water, or copper strip electric heat conductors.

Piped systems are normally 12 mm nominal bore and looped at about 250 to 300 mm centres as shown in Fig. 7.8. Closer spacing is

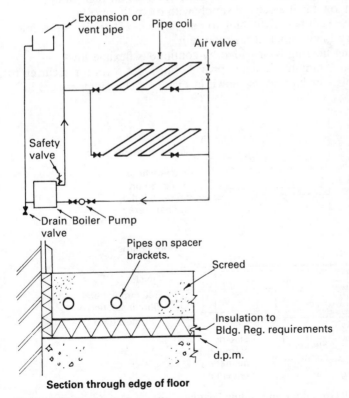

Fig. 7.8 Embedded pipe coils

possible if a high heat output is required from the wall or ceiling. To avoid discomfort, the maximum recommended surface temperatures are 49 °C for ceilings, 43 °C for walls and 29 °C at floor level. With the lower operating temperature at floor level and the tendency for heat to dissipate more easily from a reduced level, embedded floor panels offer the greatest advantage provided sufficient insulation is installed below the pipe coils. Fig. 7.8 illustrates a section through a typical solid floor with perimeter and underfloor insulation. Surface finishes are limited, as any form of carpeting will act as an insulating heat barrier. Also with a floor temperature as low as the upper twenties centigrade, the mean water temperature need not exceed 50 °C. This offers considerable fuel saving potential, particularly if supplemented with alternative energy sources such as solar panels or heat pumps.

Electric fabric heating

Floor warming by electricity involves installation of elements within the floor screed as shown in Fig. 7.9. The floor slab must be completely insulated, including the perimeter, to current Building Regulation requirements. This system operates on night time, restricted low-tariff electrical supply with room thermostat temperature control. Switching to normal-tariff electricity is possible when very low temperatures require it.

Ceiling heating is provided by continuous flexible low-temperature radiant elements provided in sheet form for securing to ceiling joists or battens as shown in Fig. 7.9. The elements are in

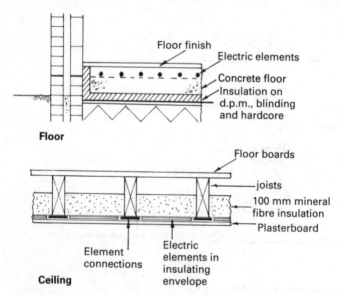

Fig. 7.9 Electric, floor and ceiling heating

parallel strips of metal in alloy foil or glass-fibre cloth impregnated
with a conducting PTFE dispersion. Supply connections are made to
'cold tails' with a normal operating temperature not exceeding 50 °C,
providing a surface temperature of about 38 °C. An insulated
backing of 100 mm mineral fibre is essential between floors, with
150 mm in roof spaces. Plasterboard of 9.5 mm or 12.7 mm provides
a suitable surface covering.

Chapter 8

Sanitation pipework to single and multiple dwellings

British Standard 5572 : 1978 is the appropriate code of practice for sanitary pipework. It provides recommendations for design, installation, testing and maintenance of sanitary pipework used in domestic, commercial and public buildings.

The two most important considerations are sufficient rate of discharge from every appliance and prevention of foul air from the drains entering a building.

The discharge and capacity data shown in Table 8.1 is necessary for design purposes. Some situations satisfying these requirements are illustrated further on.

Maximum flow rates are very important design criteria, as discharge in excess of these figures may cause siphonage of water seals from traps, hence the importance of branch pipe gradients and diameters. The depth of water seal in a trap should be 75 mm except with waste pipes of over 50 mm bore, when the depth of seal may be 50 mm. Another exception occurs at ground level where it is convenient to discharge the waste from a sink or washing machine into an external gully. Here, because of the open waste, trailing off occurs to prevent seal loss and a 38 mm water seal is sufficient.

Gradients and branch lengths

Gradients to branch pipes should be uniform and not too low to prevent self-cleansing. Practical considerations may limit the fall to about 1° or 1½°, and at the other extreme 5° to 6° must be considered the maximum to avoid self-siphonage. Fig. 8.1 shows the design curve and maximum gradient for basin wastes, and Fig. 8.2 the

Table 8.1

Appliance	Capacity (litres)	Branch pipe nominal bore (mm)	Maximum flow (litres)
WC	9	100	2.3
Urinal	4.5	40	0.15
Basin	6	32	0.6
Sink	23	40	0.9
Bath	80	40	1.1
Washing machine	180	40	0.7
Shower	—	40	0.1

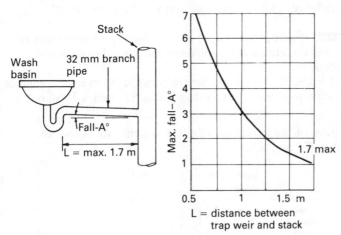

Fig. 8.1 Basin waste limitations

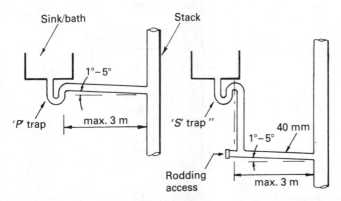

Fig. 8.2 Sink and bath waste limitations

variations for baths and sinks. Branch lengths must be as short as possible to avoid self siphonage and accumulation of deposits. The exception is large diameter WC branches which are unlikely to siphon or block but will appear unsightly if not coupled close to the stack.

Systems of sanitation

Classification of soil and waste discharge systems relates to the disposition of sanitary appliances and the method of retaining the water seal in traps. The four systems are:

1. Fully-ventilated system, used where there are a large number of appliances in ranges or widely dispersed, and where it is impractical to provide more than one wet stack.
2. Ventilated-stack system, used where close grouping of appliances around the stack permits omission of individual appliance vent pipes. Seals are retained by cross venting the stack or WC branch.
3. Single-stack system, used where close grouping of appliances is possible and the number of appliances are limited to domestic requirements.
4. Modified single-stack system, used where the occasional appliance has a branch pipe length exceeding the recommended without a vent.

Illustration of the principles of these systems are shown in Figs. 8.3, 8.4, 8.5 and 8.6.

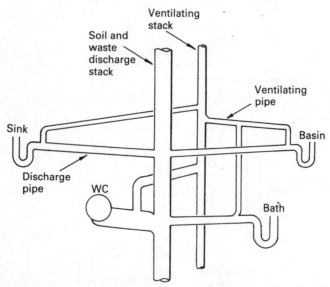

Fig. 8.3 Fully ventilated system

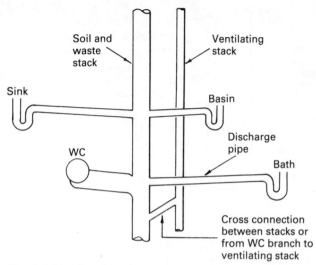

Fig. 8.4 Ventilated stack system

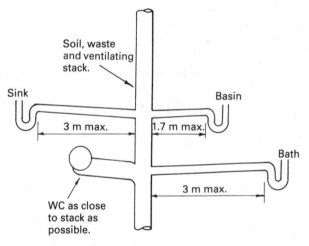

Fig. 8.5 Single-stack system

Branch connections to the stack

To ease the flow between a branch pipe and the stack, swept branches may be provided. The radius of bend is unlikely to affect the performance of bath and sink discharges but the basin branch connection must not exceed 25 mm radius. Also, basin branch connections using a $92\frac{1}{2}°$ boss must provide no more than 250 mm of change in gradient. WC branch pipes should be swept in the

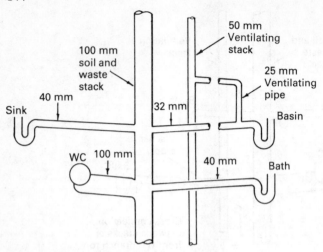

Fig. 8.6 Modified single-stack system

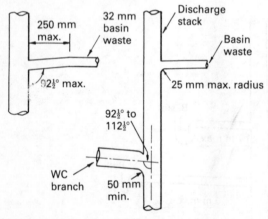

Fig. 8.7 Basin and WC branch pipe connections to the discharge stack

direction of flow with a minimum radius of 50 mm. Each of these requirements is illustrated in Fig. 8.7.

Prevention of cross flow

Cross flow refers to the discharge from one appliance backing up the branch pipe of another. In most installations this will refer to opposing WC and bath pipes at floor level. To avoid cross flow the bath branch must connect to the stack so that its centre line coincides with the stack centre line at or above the coincidence of the WC branch centre line and stack centre line. Alternatively, the

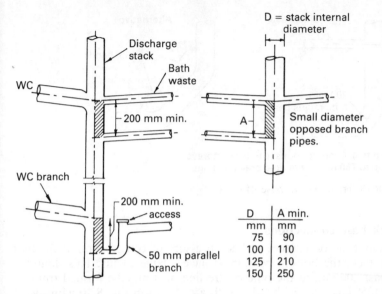

Fig. 8.8 Opposing connections

bath branch may connect at least 200 mm below WC and stack
centre lines. This is simply presented in Fig. 8.8 with the
requirements where waste pipes oppose.

'Collar boss' system

The collar boss overcomes the restrictions and difficulties of
opposing connections between soil and waste pipes at floor level.
The boss is an annular chamber permitting bath pipework to meet
the stack at any point around its circumference, without impeding
the flow of water. Fig. 8.9 shows a sectional detail of this fitting with
a diagram of its installation.

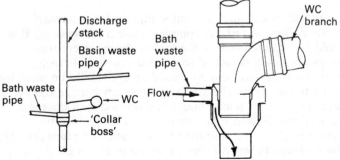

Fig. 8.9 'Collar boss' connections

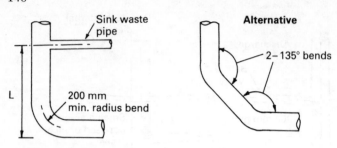

L = up to 450 mm for houses to three storeys
and up to 750 mm for multi-storey buildings

Fig. 8.10 Treatment at base of stack

Stack base connections

To avoid the occurrence of back pressure or compression at the base
of a stack, the bottom bend must have a large radius. This should be
at least 200 mm to the pipe centre line or it may be formed from
two 135° large radius bends. Both are shown in Fig. 8.10 with the
minimum dimensions between the drain invert and the lowest branch
centre line.

Further requirements

Offsets may be necessary to avoid structural obstacles. They disturb
flow conditions and may contribute to seal loss in nearby traps.
Where they are required, they must be of large radius (min.
200 mm) bends and no connections may be made to them. Where a
ventilation stack is provided with the system the offset should be
cross vented to preserve uniform pressures in the manner shown in
Fig. 8.11. Offsets above the highest branch in the dry part of a stack
are more acceptable and will not require venting. The termination or
top of the stack should be provided with a perforated dome and
arranged to be no less than 900 mm above the highest window
opening within a horizontal distance of 3 m. This is also shown in
Fig. 8.11.

Access to soil, waste and ventilating pipework is necessary
following installation, to test the complete system for leakage. It is
also required for maintenance of the system during operation.
Modern traps are easily detached to provide access to branch
pipework; if this is not convenient a rodding eye should be provided
at the branch termination. The structure must not impede the use of
access points and service ducts should be provided with doors or
removable sections. Nuisance and damage can be avoided by
locating accesses above the spill-over level of affected pipework. This
is normally mid-way between floors as shown in Fig. 8.12.

Systems vary depending on the height of the building and the

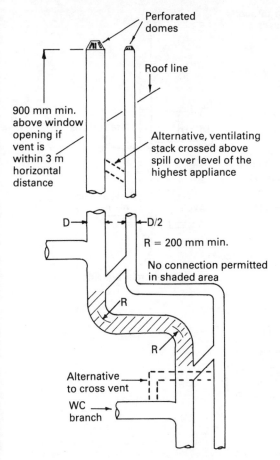

Fig. 8.11 Stack offsets and termination

number and disposition of appliances. The following illustrations indicate possible installation principles to several different situations.

Testing

Building Regulation N4 (4) section (e) requires all soil, waste and ventilating pipes '... to be capable of withstanding a smoke or air test for a minimum period of three minutes at a pressure equivalent to a head of not less than 38 mm of water.'

An air test is the most satisfactory method of determining the soundness of a system of sanitation. All water seals in traps are fully charged and test plugs inserted in the top of the vent pipes and at the base of the stack. If access is not provided here, the upper connection to the highest inspection chamber must be plugged. The most convenient test plug is fitted with a flexible tube connecting to

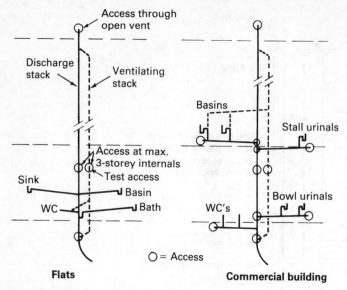

Fig. 8.12 Location of access points

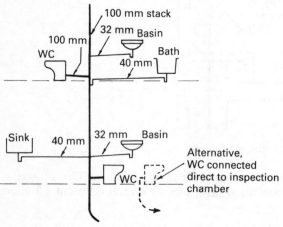

Fig. 8.13 Two-storey house

a manometer and hand bellows or bicycle pump as shown in Fig. 8.17. Alternatively, the flexible tube may be passed through the water seal of a WC pan. Air is pumped into the system until a 38 mm water gauge is achieved. The air inlet cock is closed and the pressure observed constant for three minutes. If the pressure falls, leakage may be detected by maintaining a pressure inside the pipe and brushing soap solution over pipes and joints until bubble formations indicate the defective area.

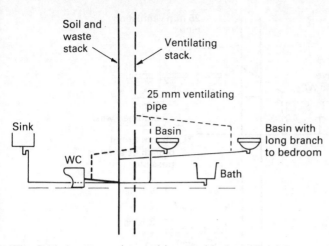

Fig. 8.14 Apartment in a multi-storey block

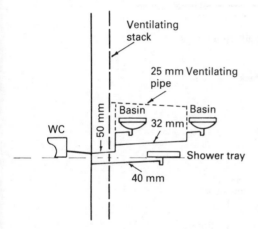

Fig. 8.15 Halls of residence

Design

A 100 mm diameter stack with branch pipe sizes previously detailed will satisfy most requirements for buildings up to six storeys. For very large or tall buildings a system based on discharge unit values representing the load-producing properties of appliances can be used to calculate the branch and stack size. These values permit a maximum flow of one quarter capacity for stacks and half capacity for branches, to prevent the effects of siphonage caused by pressure variations.

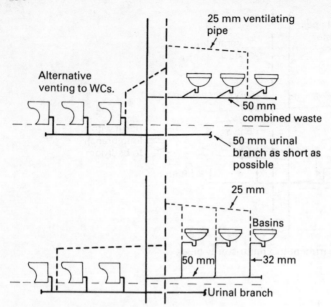

Fig. 8.16 Commercial and public buildings

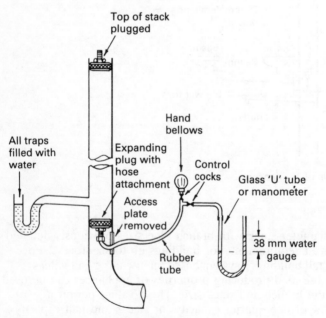

Fig. 8.17 Testing stack for leakage

Table 8.2 Discharge unit values – BS 5572 : 1978

Appliance	Use	Discharge unit value
WC	Peak domestic	7
	Peak commercial	14
	Congested	28
Basin	Peak domestic	1
	Peak commercial	3
	Congested	6
Bath	Domestic	7
	Commercial	18
Sink	Peak domestic	6
	Peak commercial	14
	Congested	27
Shower	Domestic	1
	Commercial	2
Urinal		0.3
Washing machine		4

Table 8.3 Number of discharge units on stacks – BS 5572 : 1978

Nominal bore (mm)	Approximate number
50	10
65	60
75	200
90	350
100	750
125	2500
150	5500

Table 8.4 Number of discharge units on branch pipes – BS 5572 : 1978

Nominal bore (mm)	Discharge units		
	Gradient		
	$\frac{1}{2}°$	$1\frac{1}{4}°$	$2\frac{1}{2}°$
32	—	1	1
40	—	2	8
50	—	10	26
65	—	35	95
75	—	100	230
90	120	230	460
100	230	430	1050
125	780	1500	3000
150	2000	3500	7500

Example

A four storey public building contains 8 WCs, 10 basins, 4 sinks and 5 urinals on each floor. Determine the size of stack assuming peak commercial use of appliances.

WCs 8 × 14	=	112	
Basins 10 × 3	=	30	
Sinks 4 × 14	=	56	
Urinals 5 × 0.3	=	1.5	
		199.5	
	×	4	storeys
		798	Discharge units

Reference to the design tables (tables 8.2, 8.3, 8.4) indicates that a 100 mm diameter stack is slightly too small (750 units), therefore a 125 mm stack (2500 units) will be adequate.

Ventilating pipes cannot be sized by the discharge unit method as they contain pressure variations, not hydraulic loadings. The following is a general guide:

Branch or stack discharge pipe	Branch or stack ventilating pipe diameter
Up to 75 mm bore	2/3 bore (25 mm min.)
Over 75 mm bore	$\frac{1}{2}$ bore

Chapter 9

Finishes: plasters, renders and screeds

Plasters

Building plasters are a product of gypsum rock deposits. They vary in quality, and composition, both of which are adjusted to suit different backgrounds. Tables 9.1, 9.2 indicate the British Standard classification and trade names of plasters appropriate to several different situations.

Type of plaster and suitability

Class A. Plaster of Paris
Very quick setting, therefore only suited to crack filling, patching and general repair work.

Class B. Retarded hemihydrate
Type (a) Suited to rendered or other sand-based backgrounds.

Type (b) Finishing plasters may be used neat or gauged with not more than 25 per cent lime for one- or two-coat work. Plasterboard finish plaster is used neat as a single coat only.

Class C. Anhydrous
Specified for final coat only. Normally used neat, but maybe gauged with up to 25 per cent lime by volume.

Table 9.1

Gypsum-based plasters	British Standard		Trade name
Class A Plaster of Paris	BS 1191 : Pt 1		CB Stucco
Class B Retarded hemihydrate Undercoats	BS 1191 : Pt 1		
browning		a1	Thistle Browning Thistle slow setting Browning Thistle Fibred
metal lathing		a2	Thistle Metal Lathing
Finish coats			
finish		b1	Thistle Finish
board finish		b2	Thistle Board Finish
Class C Anhydrous Finish coat	BS 1191 : Pt 1		Sirapite
Class D Keene's	BS 1191 : Pt 1		Standard Keene's Fine, Polar White Cement Standard, Polar White Cement
Pre-mixed lightweight Undercoats	BS 1191 : Pt 2		
browning		a1	Carlite Browning Carlite Browning (HSB)
metal lathing		a2	Carlite Metal Lathing
bonding		a3	Carlite Bonding Carlite Welterweight Bonding
Finish coats			
finish		b1	Carlite Finish Limelight Finishing

Class D. Keene's
Used neat for a hard final coat. Suitable where the surface may be subject to knocks or abrasion.

Special plasters
X-ray plaster. Primarily for use in hospitals and laboratories to provide protection against X-rays. The plaster is composed of a cement or retarded hemihydrate gypsum binder with a barium sulphate aggregate.

Table 9.2 Non-BS classified plasters

Gypsum plasters	Trade name
Thin coat	LSM Special Finishing
Acoustic	Thistle Acoustic
Projection (machine application)	Thistle Projection
X-ray	
Undercoat	Barytite Rough
	Barytite Fibred
Finish	Barytite Finish
Cement plasters	
Pre-mixed undercoat	Limelite Backing
Resinous (suitable squash courts)	
Undercoat	Proderite Formula 'S' Base Screed
Finish	Proderite Formula 'S' Finish

Thin-wall plasters. These are produced as a paste composed of finely ground minerals with organic binders, delivered on site in steel drums. Cement or gypsum plasters are set by chemical reaction with water, but these plasters remain moist until applied in very thin coats, hardening by drying. They are particularly suited to very large panels of concrete and may be applied by spraying.

Thermal insulating plasters. Insulating plaster is a cement- or gypsum-based material mixed with a low conductivity aggregate. The efficiency depends on the ratio of aggregate to binder; if too high, strength is impaired and, if too low, the aggregate will be less effective if it is filled with the binder. To some extent these plasters will absorb condensation from the atmosphere due to their open-textured nature. This property is of limited value and these plasters cannot be regarded as particularly significant for use where condensation is a problem.

Acoustic plaster. Plasters specified for their sound insulative properties are basically retarded hemihydrate gypsum plaster (class B) with a pumice aggregate additive. Acoustic finishing plasters should be applied to a well-keyed undercoat selected for the appropriate background and finishing conditions. Surface treatments should be avoided as they will obscure the insulative effect.

Mechanical or projection plaster. Mechanical methods of plastering have been developed in an attempt to reduce the physical effort factor required for wet finishes. This increases the time available for the skilled finishing process, creating greater efficiency in the use of skilled labour. The plaster is a mix of retarded hemihydrate plaster (class B) and anhydrous plaster (class C) providing one coat work of 12 mm thickness to walls and 8–10 mm for ceilings and other soffit areas. The plaster is fed through a hopper, mixed with water and pumped via a flexible hose to a spreader nozzle. It is levelled with a darby, followed by wood floating and final finishing with a steel float. This technique has been particularly successful when applied to suspended expanded metal lath ceilings and EML faced stud partitions.

Background

The system of plastering will depend considerably on the properties of the background material, including the level of the wall and the suction effect. Walls with considerable variations in either or both will require a three-coat work to establish a uniform finish, whereas level backgrounds with uniform suction such as dense concrete or plasterboard will require one coat only. The strength of the background is also an important factor determining the type of undercoat mix. If the undercoat is cement-based it must not be stronger than the background otherwise shrinkage will be excessive.

The background porosity and suction affect both strength and adhesion of plaster-based undercoats. By absorbing some workability of the water content of undercoats, high suction backgrounds increase the strength of the undercoat, so the proportion of sand in browning plaster undercoats can be relatively high.

Table 9.3 is provided as a guide to plaster- and cement-based applications on different backgrounds.

Table 9.3

Background	Undercoat			Finish
	Premixed lightweight plaster	Sanded browning	Cement, lime and sand	
1	Bonding	1 : 1½	1 : 1 : 6	Lightweight Class B or C
2	Browning	1 : 2½	1 : 1 : 6	Lightweight Class B or C
3	Browning or browning HSB	1 : 2	1 : 1 : 6	Lightweight Class B or C
4	Browning	1 : 2	1 : 1 : 6	Lightweight Class B or C
5	Metal lathing	1 : 2		Lightweight Class B or C
6	Metal lathing	1 : 1½ 1 : 2	1 : 1 : 6	Lightweight Class B or C
7	Bonding			Lightweight Board finish direct
8	Bonding	1 : 1½		Lightweight Class B or C

Background materials

1. Engineering bricks, concrete, concrete bricks, high density concrete blocks and closed surface lightweight concrete blocks.
2. Common bricks, calcium silicate bricks and medium density concrete.
3. Open surface lightweight aggregate concrete, weak and soft bricks.
4. No fine aggregate concrete.
5. Wood wool slabs.
6. Expanded metal lathing or welded wire mesh.
7. Plasterboard.
8. Painted or glazed surfaces treated with a bonding agent, e.g. emulsions of PVAC or bituminous solutions.

Renders

External rendering should possess a combination of properties, including durability, pleasing appearance and minimal maintenance. Modern renders are cement-based to bind the aggregates of lime and sand and to provide a moderately impervious composition. Inclusion of too much cement in the mix in order to achieve dense waterproof qualities encourages shrinkage cracking, ruining the appearance and permitting transference of water. Furthermore, surface treatments which are trowelled smooth to encourage rapid run off of water invariably weather unevenly, encouraging patched areas of dampness and streaked appearance. Recent developments indicate that rendered finishes are more successful if they are of a slightly porous composition with a textured surface treatment. Lime as the basis of a rendering mix is important to provide workability and sufficient porosity to allow easy drying after exposure to rainfall. This prevents entrapment of water behind the render which could suffer from frost action and transfer into the building subsequently causing the render to spall.

The mix is prepared by combining hydrated lime with sand to produce 'coarse-stuff'. Cement is added and mixing continued until a uniform appearance is achieved; sufficient water is finally included with the mixing process to create a workable consistency. Machine mixing is preferred to ensure a greater efficiency and more thoroughness of mix. Composition of mix depends on the degree of exposure and the background material. Tables 9.4 and 9.5 may be used as a guide.

Background preparation

It is essential that background materials are clean and free of algae and fungal growths. Loose material and other deposits must be

Table 9.4

Mix code	Cement : lime : sand	Masonry cement : sand
A	1 : $\frac{1}{4}$: 3	Not suitable
B	1 : $\frac{1}{2}$: 4	1 : 3
C	1 : 1 : 5	1 : $4\frac{1}{2}$
D	1 : 2 : 8	1 : 6

Table 9.5

Background	Surface treatment	Undercoats	Finish
1	Floated	B or C	C D if
	Textured	B or C	C sheltered
	Rough cast	A or B	B
	Pebble dash	A or B	B
2	Floated	B or C	C
	Textured	C	C
	Rough cast	B	B
	Pebble dash	B	B
3	Floated	C	C
	Textured	C	C
	Rough cast	Not advisable on weak	
	Pebble dash	backgrounds as these treatments should be strong mixes.	
4	Floated	C	C
	Textured	C D if sheltered	C D if sheltered
	Rough cast	A or B	B
	Pebble dash	A or B	B
5	Floated	C or D	D
	Textured	C or D	D
	Rough cast	Not advisable	
	Pebble dash	Not advisable	
6	Floated	B or C	C
	Textured	B or C	C
	Rough cast	A or B	B
	Pebble dash	A or B	B

brushed from the surface to avoid inhibiting the bonding effect of the undercoat. The bonding or undercoat should not exceed 13 mm and subsequent coats about 10 mm. To control the absorbing effects of the background, the surface should be dampened, but not to excess, otherwise poor bonding and slumping during application will result. The surface of undercoats should be scratched to provide a key for subsequent coats, and suitable protection must be provided when exposed to hot sun or frost. Damp cloths or light spraying are adequate during hot weather to prevent premature drying and cracking.

Concrete

Treatment of concrete surfaces to receive rendered finishes may include modifications to the formwork or post treatment of the finished concrete. Pretreatment includes the use of the following:

1. Surface retarders painted on the inside of formwork. When the formwork is struck, all loose materials must be washed away with water and stiff brushing.
2. Expanded metal lathing or wire mesh attached to the inside of formwork. As the formwork is struck, the concrete surface is disturbed, leaving a rough key.
3. Preformed fibre or rubber mouldings secured to the formwork to leave an indented key over the concrete surface.

   Post-treatments should always be preceded by thorough washing of the surface to remove laitance, efflorescence and mould oil. The following methods may be considered:

1. Hacking or scarifying by bush hammer or abrasive blast process. This traditional technique is only effective if an overall 3 mm is removed to expose the aggregate.
2. Spatter dash. A strong mix (1 : 2) of cement and sharp sand is mixed to make a slurry. After cleaning and dampening the concrete, the slurry is trowelled on to create an uneven keyed finish for subsequent applications. Alternatively, a PVA emulsion bonding agent may be added to the water before mixing with the cement and aggregate. Thin solutions are brushed on the surface and stippled with a stiff brush, and thicker solutions are applied by trowel and scored to provide a ribbed key.
3. Other proprietary bonding agents exist for mixing with cement to form a thin slurry which is scrubbed onto the surface just prior to applying the undercoat.
4. Galvanised expanded metal lathing or welded wire mesh fixed at a maximum of 600 mm spacing with shot-fired nails and washers or by ordinary galvanised nails located with timber blocks cast into the concrete. To preserve uniformity, the strands of expanded metal should run in one direction only; on walls, they should slope inwards and downwards against the concrete face to maximise the keying effect. Overlap between sheets should never be less than 25 mm and wired together at a maximum of 150 mm intervals.

Bricks and blocks

Calcium silicate bricks and low strength concrete blocks frequently have a surface texture suited for direct application of undercoats.

Clay bricks are less suited, and during construction joints should be raked out at least 12 mm to provide a key. Old brickwork with soft mortar joints should be similarly treated after brushing impurities away. If the mortar is hard or the surface painted, metal lathing may be attached to the surface. Alternatively, the surface must be thoroughly hacked to remove the paint and to provide a key.

Lightweight backgrounds

Woodwool slabs and other less dense backgrounds should not be rendered directly. Ideally the surface should be pretreated with galvanised wire mesh reinforcement secured with staples at a maximum of 300 mm spacing. Three coats are preferred, the first in the form of spatterdash to bond subsequent treatment.

Finishes

Finishes to external renders may be:
(a) Smooth, float finish.
(b) Textured, scraped or machine-applied.
(c) Rough cast or hand thrown.
(d) Pebble or dry dash.

Smooth

These finishes are either steel-trowelled or wood-floated. They are no longer recommended as the degree of workmanship required to achieve a crack- and discolouration-free finish is difficult. Excessive use of a steel trowel draws cement to the surface, and unless carefully controlled drying is possible the surface will craze.

Textured

Textured finishes are achieved by treating the final coat shortly after application, but not after setting has commenced. Many varieties of attractive finish are possible, mainly recessed and keyed by hand tools. Machine-applied finishes are sprayed or thrown particles to provide a decorative textured treatment. Both hand and machine finishes are less likely to crack or craze, and the occurrence of small cracking will not be detected. Although the imperfect nature of the surface will encourage the entrapment of dirt this will not be obvious due to the surface irregularity.

Rough cast

A wet mixture of aggregate and cement is thrown onto the undercoat to achieve an irregular finish with properties similar to textured surfaces. These are particularly suited to severely exposed conditions.

Pebble dash

This is sometimes known as dry dash or dry cast as the aggregate is thrown on dry to a wet background. It is a very effective cover, masking cracking and other superficial irregularities and has very good weathering properties.

Table 9.6

Application	Thickness
Spatterdash	8–16 mm
Undercoat	8–13 mm
Smooth finish	6–10 mm
Textured finish	10–13 mm (about 3 mm surface treated)
Dry/pebble dash	10 mm

Floor screeds

Screed finishing is traditionally the plasterer's trade. Contemporary power float or power trowel methods provide an excellent finish to concrete dispensing with the need for additional time-consuming treatment. Nevertheless, screeds are still required either for particularly fine floor finishes, finishes laid to a fall, or to cover thermal insulation and to accommodate services and underfloor warming elements.

Screeds are normally composed of a mixture of cement and fine aggregate. These are known as dense screeds. Improvements in insulation standards often require the use of cellular aggregates to achieve lightweight screeds; these are manufactured from vermiculite, expanded clay, shale or pulverised fuel ash.

Typical specifications

Fine screed
Mix proportions by weight, 1 part of Portland cement to 3 parts fine aggregate (up to $4\frac{1}{2}$ parts fine aggregate are possible).

Thick screed
Mix proportions, 1 part of Portland cement to $1\frac{1}{2}$ parts of fine aggregate and 3 parts coarse aggregate up to 10 mm.

Water. Sufficient to permit thorough compaction.

Maximum bay size 15 m^2 on hardened base and 30 m^2 if laid monolithically. Ratio of bay sides to be as close to 1 : $1\frac{1}{2}$ as possible.
Thickness. On hardened bases a minimum of 40 mm is acceptable. If the base is hacked, cleaned and grouted, 12 mm is possible although a thicker screed is always preferable.

On unset bases (monolithic construction) complete bonding may be obtained and a 12 mm thickness is adequate. Where it is impossible to achieve a bond on mastic or integrated thermal insulation materials, 50 mm is regarded as the minimum acceptable screed thickness. If the base material is compressible, 65 mm must be specified.

Curing 3 to 7 days by covering with polythene. After curing, the screed should be allowed to dry thoroughly before application of the floor finish. An exact period is difficult to determine, as drying time will vary depending on the heating and ventilation of the building and the screed thickness. As a guide, the rule of thumb of one month for every 25 mm screed thickness will provide some indication. More accurate determination of screed dryness may be obtained by hygrometer test. This is detailed in the *Building Research Establishment Digest* No. 18 (second series).

Lightweight aggregate screeds

These have the advantage of improving the thermal insulation properties of roofs and floors. They also accommodate services with an insignificant increase in weight for the increase in height or cover required. However, the thickness must never be less than 40 mm to provide sufficient strength, and where point loads are likely a 10 mm topping of cement and sand (1 : 4) must be laid monolithically with the insulating screed. They are not to be used with underfloor heating as their insulative value will prevent distribution of heat. Vermiculite mixed with cement in the ratio of 6 : 1 or expanded clay, shale or pulverised fuel ash in the ratio of 10 : 1 are commonly specified lightweight screeds.

Roof screeds

Screeding of dense concrete flat roofs is often required to provide sufficient thermal insulation. Lightweight screeds of foamed slag, pulverised fuel ash and expanded clay mixed with cement in the ratio of 8–10 : 1 are often recommended. Lower density screeds can be produced from exfoliated vermiculite or perlite in the ratio of 5 : 1 cement. The thickness, in similarity with other lightweight screed applications, should not be less than 40 mm. For roof insulation it is likely to be considerably more, with a monolithic finish of cement and sand to receive the surface treatment. Complete drying of the screed is unlikely in the UK as the average rainfall exceeds the expected rate of evaporation for most of the year. When the waterproof covering is applied, entrapped water will escape very slowly, encouraging condensation, high heat loss and staining of the internal finishes. To prevent this, roof ventilation of the type shown

163

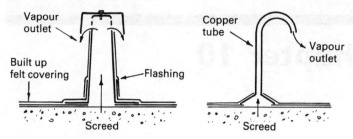

Fig. 9.1 Screed ventilators

in Fig. 9.1 should be provided every 20 m² or at 6 m intervals, and
the base layer of felt spot or partially bonded to encourage
evaporation.

Condensation within the structural roof can produce the same
effect as trapped water, but can be prevented with a vapour barrier
of bituminous felt provided over the structural concrete and under
the lightweight screed. Alternatively a vapour check may be
provided at ceiling level. This can be foil-backed plasterboard or
polythene with joints sealed with tape or sealing compound. Both
possibilities are shown in Fig. 9.2.

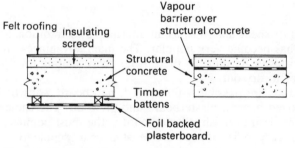

Fig. 9.2 Vapour barriers to concrete roofs

Chapter 10

Dry-lining techniques and floating-floor construction

Dry lining

The use of board or sheet materials as an alternative to traditional wet applications has become very popular. The main advantages are simplification and faster finishing as there is no delay waiting for render and plaster to dry out.

Several materials are possible – fibreboard, hardboard, softwood tongued and grooved boards, plasterboard and plywood. The choice will depend on the desired finish. Plasterboard is the most popular, being economical, easy to fix, of good fire resistance and simple to finish by emulsion painting, skim plastering or papering.

Plasterboard consists of a core of gypsum plaster set between and bonded to two sheets of paper. British Standards 1230 provides five types of plasterboard:

1. *Gypsum wallboard*. This has one side with an ivory-coloured paper suitable for direct decoration with paint and the other side is a double layer of grey paper for board plaster application. The edges are square, rounded, bevelled or tapered.
2. *Gypsum lath*. This is a narrow width plasterboard with rounded edges designed to receive board finish plaster on one side only.
3. *Gypsum baseboard*. This is available in wider sheets than laths and may be plastered either side. Edges are square.
4. *Gypsum plank*. This is available with two grey surfaces for plaster finishing or with one ivory face for direct decoration. The former has square edges and the latter tapered for taped flush joints.

Both are produced in narrow widths and greater thickness than other plasterboards and are frequently specified for linings and fire-protective casings for structural members.

5. *Insulating gypsum plasterboard.* All types of plasterboard are available with aluminium foil backing to improve their thermal-insulating properties.

Table 10.1 BS Board sizes

Type	Thickness (mm)	Width (mm)	Length (mm)
Wallboard	9.5 and 12.7	600, 900, 1200	1800, 2350, 2400, 2700, 3000
Lath	9.5 and 12.7	406	1200
Baseboard	9.5	914	1200
Plank	19	600	2350, 2400, 2700, 3000

Note: Wallboard and plank are produced 2.35 m long to suit Building Regulation floor to ceiling height of 2.30 m min

Other types of plasterboard are also available to meet specific requirements. Some examples include vapour check wallboard, thermal board and fire protection board. Vapour check wallboard has an aerated gypsum core with ivory-coloured surface and tapered joints on the outer side, with a layer of blue polythene film bonded to the back. It functions as both vapour barrier and dry lining, reducing the effect of interstitial condensation. Thermal board is basically gypsum wallboard with a layer of expanded polystyrene bonded to its surface. It is available in various overall thicknesses from 22 mm to 65 mm and combines the advantages of a drylining with improved thermal insulation for new buildings together with upgrading of existing buildings. Improved fire resistance is possible by using a gypsum plasterboard with the addition of glass fibre and vermiculite to the core material. This is an ideal dry lining where high standards of fire resistance are required. It is also available with foil backing to improve its insulating properties. Boards normally specified for dry lining are provided with a tapered edge. For treatment at corners, a square edge is often preferred with a bevelled edge being more suitable where the joints are featured. Each are illustrated in Fig. 10.1.

Plasterboard fixing
Plasterboard may be secured by nailing to a timber frame, with plasterbonding to solid backgrounds or by fixing to metal channels secured to brick or block walls.

Timber-framed support
Timber framing may be the studwork to the inner leaf of a timber-

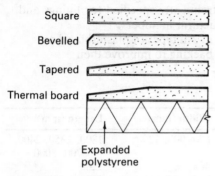

Square

Bevelled

Tapered

Thermal board

Expanded
polystyrene

Fig. 10.1 Plasterboard edge finishes

framed external wall, a studwork party wall or a timber stud
partition. Alternatively it could be 38 mm × 19 mm battening shot-
fired or screwed to a plugged solid background of bricks or blocks.
With this technique the wall must be truly plumb otherwise the
battening will require packing out with spacers. The studwork or
battening should have sufficient dimensions so that the plasterboard
can be nailed at least 13 mm from the end and 10 mm from any
paper-bound edges. The batten or stud spacing will depend on the
board width and thickness, and noggins need only be provided for
rigidity of the studs and where horizontal fixing of boards is
specified. Horizontal fixing should be avoided, except at the top and
the bottom, as plasterboard is manufactured to suit storey heights.

Table 10.2

Board thickness (mm)	Board width (mm)	Stud or batten centres (mm)
9.5	900	450
	1200	400
12.7	600	600
	900	450
19	1200	600
	600	600

Figure 10.2 Illustrates the method of securing 900 mm wide wallboard to
battened blockwork.

Nailing

Nailing to supports is provided at 150 mm centres with 2 mm
galvanised taper-headed nails of 30 mm length for 9.5 mm boards
and 40 mm length for 12.7 mm boards. Nails should be driven firmly
until the head leaves a shallow depression in the board. This is
smoothed by spot plastering before commencement of decoration.
For 19 mm plasterboard and the outer skin of two-layer work,

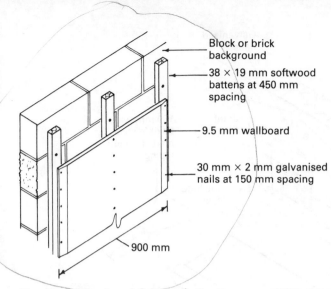

Fig. 10.2 Plasterboard fixing to battens

60 mm × 2.6 mm galvanised nails are recommended. Clout head galvanised nails are preferred for use with expanded polystyrene-backed plasterboard. Care must be taken to ensure that the polystyrene backing is not compressed by nailing too hard.

Plasterboard jointing

Tapered edge boards for dry lining may be jointed and finished by manual or mechanical means. Various proprietary implements exist for automatic joint taping and finishing as well as corner treatment and nail spotting. This section is concerned only with manual methods. Hand tools are illustrated with complete treatment of a tapered joint in Fig. 10.3. Initially, the gap and trough between the boards are filled with a gypsum filler. Whilst this is moist, 53 mm wide joint reinforcing tape is pressed into the filler to bind the two adjacent boards. A new layer of filler is applied over the tape to level the joint with the plasterboard surface. Finally, a very fine finishing compound is mixed with water and applied over a width of 200 mm to compensate for any surface irregularity. Two layers of finishing, the second 250 mm wide, may be necessary. Any imperfections or projections at the edge of the finishing treatment can be feathered out with a damp sponge.

Treatment at cut edges and corners

Joints between cut edges and tapered edges are unavoidable. Where these occur the joint is filled, and after drying a thin layer of finishing compound is used to adhere jointing tape to the surface. Final treatment is the same as between two taper-edge boards.

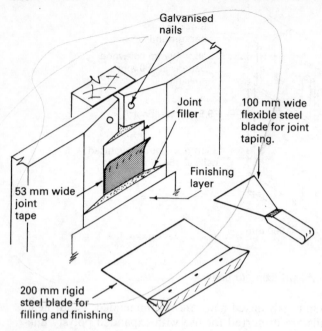

Galvanised nails

Joint filler

100 mm wide flexible steel blade for joint taping.

Finishing layer

53 mm wide joint tape

200 mm rigid steel blade for filling and finishing

Fig. 10.3 Plasterboard joint treatment

Internal corners are treated similarly with 75 to 100 mm of joint finish applied over each side of the corner. External corners are treated differently. Here, each surface is pretreated with a 50 mm-wide skim of joint filler followed by an application of a steel strip-reinforced corner tape. This provides an accurate, robust corner treatment which is completed with a thin layer of 125 mm-wide joint finishing to each side of the angle. Treatment at corners and edges is illustrated in Fig. 10.4.

Plasterbonding

Plasterbonding is a technique for applying plasterboard to solid backgrounds of brick or concrete blocks. Small pads about 100 mm × 75 mm, of bitumen-impregnated fibreboard are stuck to the wall with a plasterwork general purpose adhesive. The arrangement is shown in Fig. 10.5, with the top pad 230 mm from the ceiling and the lower pad 100 mm from the floor. Intermediate pads are placed at a maximum of 1.070 m vertically and 450 mm horizontally to suit 900 mm-wide wallboard. The pads are aligned with a spirit level and allowed up to two hours to dry. In between the pads, dabs of bonding coat or board finish plaster are applied with a trowel. Each dab is the length of the trowel and deep enough to project beyond the pads. A gap of 50 to 75 mm should remain between each dab and about 25 mm allowed before edge joints. Only sufficient dabs for one board should be applied to the wall. The board is cut about

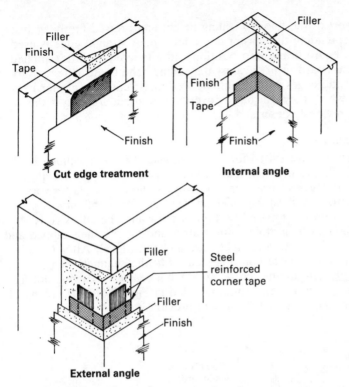

Fig. 10.4 Square edge and corner treatment

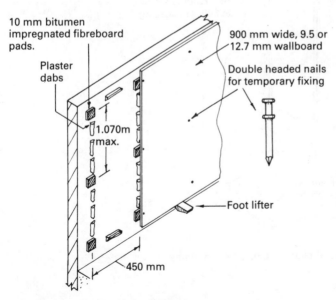

Fig. 10.5 Plasterboard dry lining

13 mm short of the floor to ceiling height and pressed firmly against the pads spreading the plaster dab adhesive. A straight edge is used to tap the board firmly into position and a tapered footlifter applied to raise the board firmly against the ceiling. The board is temporarily located at the edges with double-headed nails, until the plaster has set, after which they can be removed and re-used. Jointing is the same as previously described.

Metal channels

This system is most suited for 1200 mm wide 12.7 mm wallboard, vapour check and thermal boards. Galvanised lightweight steel channels are stuck to brick or blockwork walls with a plasterwork general purpose adhesive at 600 mm centres with a 150 mm stop section close to the top and bottom of the wall. The adhesive is applied 200 mm long at 450 mm centres and the channels tapped and plumbed vertical with a straight edge and spirit level. After the adhesive has set, plasterboard is secured to the channels using countersunk self-tapping screws at 300 mm centres occurring not less than 10 mm from edges. This is detailed in Fig. 10.6 and finishing of joints is the same as that illustrated previously.

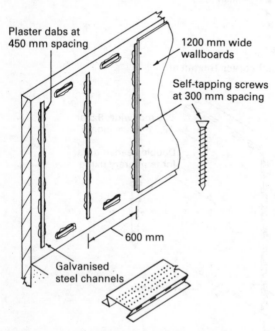

Fig. 10.6 Metal channel fixing for dry lining

Floating floor construction

Part G of the Building Regulations requires floors which separate
dwellings, as in flats and maisonettes, to have a satisfactory standard
of resistance to airborne and impact sound transmission. The
performance of a floor must compare satisfactorily with sixteen
octave bands in the frequency range 100 Hz to 3150 Hz. This is
shown graphically in Fig. 10.7 for airborne sound insulation and
Fig. 10.8 for impact sound insulation.

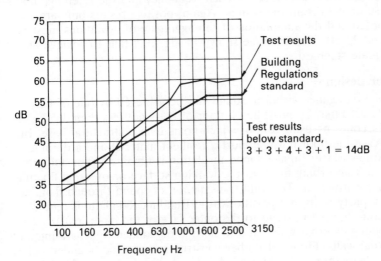

Fig. 10.7 Airborne sound insulation

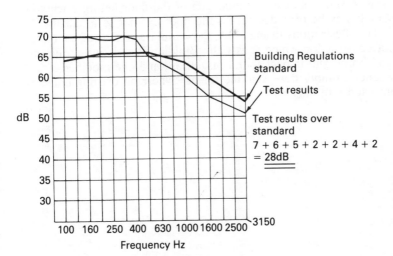

Fig. 10.8 Impact sound transmission

Superimposed on the graph are representative test results to illustrate the exact requirements of the Building Regulations. For airborne sound the performance is calculated by averaging the measured values at each frequency and subtracting these figures from the mandatory reference. All positive values are summed, negative values are ignored, as shown on the graph. This aggregate of deviations must not exceed 23 dB, so the example shown in Fig. 10.7 is satisfactory. The performance for impact sound transmittance is averaged for each frequency and the reference limits are subtracted from the results. Again, positive results only are recorded and the summation of the aggregate of deviations must not exceed 23 dB. The example shown in Fig. 10.8 fails to achieve an adequate standard.

Floor design

Insulation against airborne sound can be achieved by one of two methods. First, by providing a floor of mass not less than 365 kg/m^2 and second, by using a floor which floats over a structural base. In this case the floor must support a resilient material and a durable walking surface. Impact sound insulation is obtained by installation of a soft absorbing finish to the walking surface or by using floating floor construction. To avoid transmission of sound through flanking walls, party walls and partitions, the Building Regulations require concrete floors to extend to the outer face of the inner leaf of adjoining external walls, and to be tied or bonded to separating and internal walls. Floors of timber construction must be bounded below on at least three sides by walls having an average mass of 415 kg/m^2, measured over any portion of the wall. Also, flanking walls must rise at least 600 mm above the underside of the floor before a window opening may be provided unless the floor extends to a balcony.

The illustrations in Fig. 10.9 show four variations of concrete floor construction which satisfy the Building Regulation requirements for airborne and impact sound insulation, and one modified suspended timber floor suitable for use where an existing house is converted into flats.

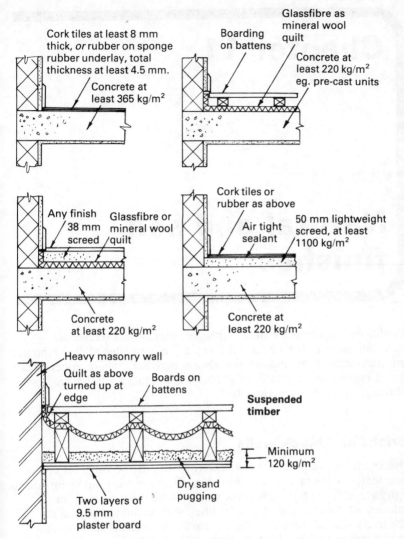

Fig. 10.9 Floor construction satisfying Part G of the Building Regulations

Chapter 11

Natural and self finishes

Traditional wet finishes such as plaster, render and screed are superimposed on the structural surface. They involve skilled labour, are time-consuming and are not always necessary where appearance is not important or where an alternative satisfactory finish may be obtained from the structure itself.

Bricks and block walls

Bricks are no longer limited to external use, as the range of colours and textures available is considerable. They are well suited for use in entrance halls and provide attractive treatment to many other interior situations. The choice of brick will be limited to one of relatively smooth texture. The regularity of concrete or sand-lime bricks finished with a flush or feature joint will provide a very uniform wall. Mortar should be 1 : 3 : 10–12 or 1 : 2 : 8–9 cement, lime and sand for spring and summer construction and 1 : 1 : 5–6 if there is a risk of frost.

Concrete blocks are also manufactured in many different colours and textures. The quality of finish varies, including machined smoothness and exposed aggregate surfaces. Exposed aggregate finishes are cast as a facing onto a lightweight concrete background whilst the block is in the mould. Profiled or sculptured surface treatment is also possible while the concrete is partly set. Glazed blocks are produced to permit construction and finishing of an interior glazed partition wall or the inner leaf of a cavity wall in one operation. A combination of thermo-setting resinous binder and glass silica sand is cast as a liquid onto the surface of the blocks to provide a 3 mm thickness. The liquid penetrates the pores of the concrete and when fired becomes an integral part of the block. Many colours are possible by varying the pigment content.

Brick and block floors

Bricks are very fashionable for floor construction, although they will require supplementary support from concrete of at least 75 mm thickness. Normal bricks of hard durable properties are most suitable and special bricks of about 50 mm thickness with no frog are produced particularly for brick paving. Both types may be laid flat or on edge in various arrangements. Herring-bone pattern is very attractive, but involves awkward cutting around the perimeter. Basket weave or coursed brickwork with breaking joints as shown in Fig. 11.1 are both simpler while still retaining a presentable finish.

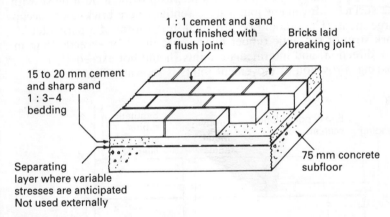

Fig. 11.1 Brick floor construction

Where variable stresses are likely, differential movement between a brick floor and a concrete sub-floor will cause superficial cracking. To prevent this, a separating layer of polythene, building paper or bituminous felt should be provided on top of the sub-floor. Alternatively a semi-dry bedding will isolate the bedded finish from the base. This is composed of cement and sharp sand in the ratio of 1 : 3½–4 with a water cement ratio of about 0.55 (28 litres of water : 1 bag of cement). After tamping and levelling the mix, a slurry of cement and fine sand in the ratio of 1 : 1 is spread in about 2 mm thickness over the surface before receiving the bricks. This technique is illustrated in Fig. 11.2.

Joints should be filled or grouted no sooner than 12 hours from laying the bricks to permit adequate setting. Grout is composed of

176

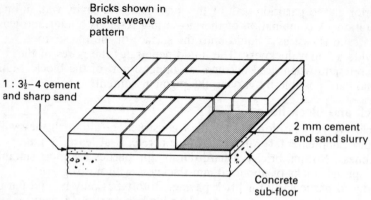

Bricks shown in
basket weave
pattern

1 : 3½–4 cement
and sharp sand

2 mm cement
and sand slurry

Concrete
sub-floor

Fig. 11.2 Semi-dry bedding of brick floors

cement and fine sand in the ratio of 1 : 1, mixed to a paste consistency and worked into the joints to provide a flush finish with the surface. Movement joints are essential where brick floors occupy large areas. They should be provided in the form of a perimeter cork strip or synthetic rubber filler where the floor exceeds 15 m in any direction, and intermediate joints should not exceed 15 m spacing. Provision of movement joints is shown in Fig. 11.3.

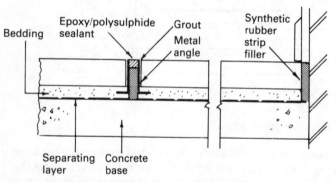

Bedding

Epoxy/polysulphide
sealant

Grout

Metal
angle

Synthetic
rubber
strip
filler

Separating
layer

Concrete
base

Fig. 11.3 Movement joints in brick floors

Concrete paving blocks and slabs

Block paving is an alternative external finish to brickwork. It provides an attractive, durable surface suitable for pedestrians and light traffic. The shape and colour of blocks may be varied to suit specific situations ranging from pavings and play areas to docks, warehouses and roads. Shapes and dimensions are indicated in Fig. 11.4 and the method of laying is shown in Fig. 11.5.

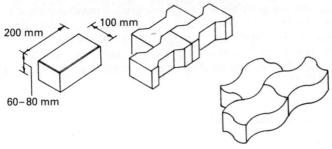

Fig. 11.4 Types of concrete floor block

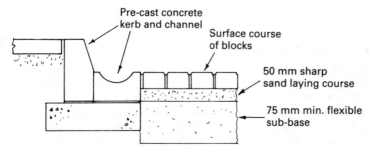

Fig. 11.5 Construction using concrete floor blocks

Sitework is undertaken in four stages. Initially, edge restraint is provided by pre-cast concrete kerbs and channels to prevent block movement. A flexible sub-base from 75–450 mm deep is created from crushed granular materials, including chalk, rock, slag or fuel ash. The depth will depend on the subsoil composition. Sandy materials will require the least superimposed thickness and clays the greatest. If the sub-base layer exceeds 200 mm it must be placed and compacted in two or more layers. An alternative sub-base may be a lean mix concrete having an aggregate of between 20 and 40 mm. Cement should be added to produce a compressive strength of between 10 and 20 N/mm^2 at 28 days. After completion of the sub-base a 50 mm laying course of sharp sand compacted by plate vibrator receives the surface course of blocks. When the blocks are positioned the area is brushed with dry sand to fill all joints before final compaction. For vehicular traffic, rectangular blocks laid in a herringbone pattern produce the best interlock and resistance to the effects of braking.

Concrete flag stones

Concrete slabs or flags are manufactured to the requirements of BS 368. They are composed of cement and crushed igneous rock.

Colour variations are possible by using coloured cements which may be applied throughout the unit or in a surface layer of not less than 12 mm thickness. Two thicknesses are available, 50 mm for pedestrian walkways and 63 mm for use where vehicular access occurs. The four British Standard sizes are represented by letters as shown in Table 11.1.

Table 11.1

Type	Size (mm)
A	600 × 450
B	600 × 600
C	600 × 750
D	600 × 900

Method of laying paving flags

The sub-grade is prepared by removing at least 100 mm of topsoil followed by thorough compaction and consolidation with a vibrating roller. If vegetable soil remains it must be treated with a proprietory weed killer or sodium chlorate in solution. Restraint from movement is provided by kerbing or other suitable edging and a flexible sub-base is prepared as for concrete blocks, from crushed stone, clinker, etc. Flags should be bedded on a dry mix of mortar (1 : 3) of approximately 25 mm thickness. The five-spot method of applying dabs of wet mortar for flag bedding is no longer recommended as this gives insufficient support. Flags are butt-jointed, providing a

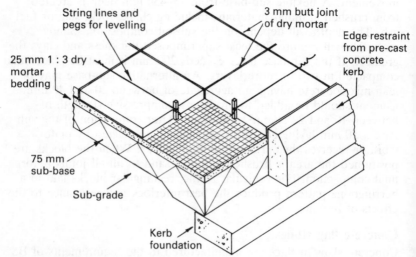

Fig. 11.6 Concrete paving flags

3 mm maximum joint and levelled from string lines suspended from timber pegs driven into the sub-base. Dry mortar is brushed over the surface to fill the joints or a 6 mm gap left and filled with a fairly stiff mortar to leave a feature joint. Laying may be with straight joints as shown in Fig. 11.6 or with a broken staggered joint. For vehicular access 63 mm flags are used on a 75 mm sub-base of concrete in the ratios of 1 : 2 : 4. Where heavy vehicles are anticipated the concrete should increase to 150 mm.

Crazy paving

Crazy paving is an effective and attractive alternative external surface treatment using broken paving slabs or quarried stone or slate. Careful selection of the slabs will minimise cutting, but some trimming is inevitable to provide joints of between 12 to 25 mm. Trimming and shaping may be undertaken by supporting the slab about 50 mm from the edge on a piece of steel angle and striking the projection with a mason's hammer or hammer and chisel.

Laying for pedestrian or vehicle use is the same as for paving flags, except that joints should be raked out well into the dry mortar bedding and filled and pointed with a fairly stiff 1 : 3 cement and sand mortar.

Wall panels

Walls may be panelled in many different materials. The most attractive are timber – based in the form of plywood or timber boarding. Where appearance is less important or an alternative finish is necessary, fibreboard or plastic-faced plywood could be employed.

Plywood

For wall panelling the standard sheets of 2.440 m height × 1.220 m width are ideal. They are produced in several thicknesses to suit various applications, 3, 4, 6.5, 9, 12, 15, 18, 21 and 24 mm. Most panels are composed of three or five laminates with the grain of each veneer running at right angles to the adjacent veneer. This tends to imbalance the construction and could cause distortion, hence the gradual introduction of four and six ply panels with the central pair of laminates arranged symmetrically.

Numerous surface treatments include:

1. *Film-faced.* A phenolic resin-impregnated paper surface, providing a hard, hygienic finish with good resistance to abrasion.
2. *Decorative film-faced.* Greater variety of colour is obtained by applying polyvinyl, polyester or acrylic resins to the paper surface.
3. *Decorative veneers.* An economic method of providing an attractive finish from unusual and expensive timber. Standard plywood panels are overlaid on the exposed surface.

4. *Plastic laminates*. Coloured plastic laminate surface adhered to a plywood base.
5. *GRP-faced*. A very hardwearing surface treatment suitable for external use. Polyester or phenolic resins reinforced with glass fibres which may be finished smooth, textured or patterned.
6. *Metal-faced*. Aluminium or copper sheet or patterned foil bonded to standard plywood. Used where special effects are desirable and for industrial doors, machinery and chutes.
7. *Rubber-faced*. Embossed rubber-bonded to plywood for industrial use and for impact sound-absorbing floor panels.
8. *Mineral aggregate finished*. Mineral chippings bonded by an epoxy resin to a plywood base. This provides a decorative rough cast surface suitable for external panels and non-slip flooring.

Many other compositions are available including asbestos, lead and cork-lining for various applications. Special requirements can be manufactured by arrangement with suppliers.

Fibre building boards

Fibre boards include hardboard, medium board and softboard, all produced to the requirements of BS 1142. They are manufactured by processing natural wood fibres into sheets and compressing them to a varying extent. Additives may be introduced to improve the properties of certain boards, but most of the fibre adhesion is obtained from properties contained in the timber. Softboard has little strength and is produced mainly for insulation purposes, where it may be sandwiched within the structure. It could be used as a ceiling material, but in many situations will require treatment to satisfy the spread of flame requirements of the Building Regulations. Hardboard is a cheaper alternative to plywood, suitable where appearance is not critical. It is available in 3.2, 4.8, 6 and 8 mm thicknesses with panel sizes up to 2.440 × 1.220 m. The surface is smooth for general use or pre-sanded and sealed for paint treatment.

Pre-decorated hardboard with plastic, enamelled or wood veneered finishes are made in plain sheets. Alternatives include simulated tiles, bricks and wood grain or plank superimposed on standard 3.2 mm board. Medium board is less dense than hardboard and is manufactured in 9 and 12 mm thicknesses. It is available undecorated or predecorated, the latter with embossed imitation brick, stone, weather boarding or planking if required.

Panels of plywood and fibreboard may be fixed direct onto sound plumb walls with adhesive or preferably by nailing to 38 × 25 mm battens at approximately 400 mm spacing. Horizontal battens should be provided at 600 mm spacing for hardboard and also for plywood panelling of 9 mm and less thickness. If a timber stud partition is the basis of support for panelling, the studding should be at least 75 mm × 50 mm at 400 mm spacing with noggins every 600 mm

Moulded timber and plywood panelling requirements

Timber-clad walls have an extensive history of use as external panelling. Simple featheredge and shiplap boarding have a long association with summerhouses, garages and other outbuildings. Contemporary structures often feature panelling under windows and as a finish to gables, but the extent of cladding is strictly limited because of the spread of fire risk. Part E7 of the Building Regulations defines the restricted use of combustible cladding, stating that no part of a timber-clad building may be within 1 m of the boundary.

At distances over approximately 5 m and below 15 m height, all the cladding may be timber. At intermediate distances the amount permitted is a proportion of the wall area, depending on the distance and the type of building. Exact calculation is determined by reference to schedule 10 of the Building Regulations. Parts B1 and B2 of the Regulations require only the use of naturally durable timbers or timber which is made durable by application of approved preservatives, namely a copper-chrome-arsenic solution which is pressure impregnated or organic solvents applied by double vacuum impregnation. These treatments are specified in schedule 5 of the Building Regulations with the acceptable species of timber and their minimum dimensions. Plywood must not be less than 9 mm thick and graded for external use. The adhesive bonding and timber quality must satisfy the weather- and boil-proof (WBP) requirements of BS 1455. Timber boards must be at least 16 mm thick and featheredge boards 16 mm minimum at the thicker edge, reducing to not less than 6 mm at the thinner edge.

Timber cladding, particularly of the tongued and grooved board form, has become very popular as an internal structural and applied finish. Part E to the Building Regulations limits the area of combustible wall surface to a maximum of half the floor surface, which leaves sufficient in most rooms for provision of some attractive timber treatment to walls. Figure 11.7 show some standard profiles suitable for internal and external applications. Featheredge and shiplap are normally only used externally, although shiplap could also be used internally.

Timber flooring

Hardwood strip and block flooring in addition to softwood board floors are considered in Chapter 14 of *Construction Processes, Level 1* by the same author. Here some variations suitable for use in sports centres and squash courts are considered. Figure 11.8 shows the method of securing hardwood-tongued and grooved-strip flooring to an underlay of softwood boarding on a screeded or power-float finished concrete. Figure 11.9 incorporates controlled resilience using rubber pads under the support joists. These are designed to reduce the physical damage by impact whilst still retaining sufficient ball response.

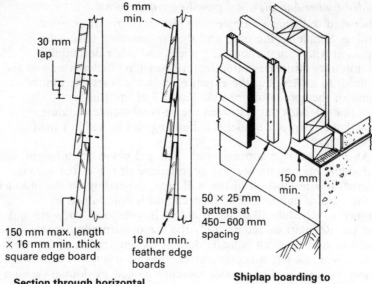

30 mm lap

6 mm min.

150 mm max. length × 16 mm min. thick square edge board

16 mm min. feather edge boards

Section through horizontal boards

50 × 25 mm battens at 450–600 mm spacing

150 mm min.

Shiplap boarding to a timber framed structure

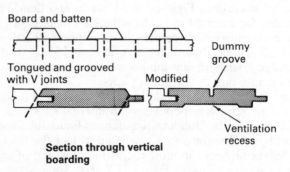

Board and batten

Tongued and grooved with V joints

Modified

Dummy groove

Ventilation recess

Section through vertical boarding

Fig. 11.7 Timber cladding

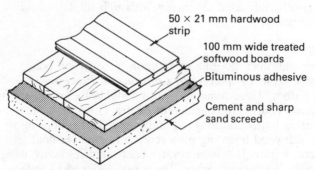

50 × 21 mm hardwood strip

100 mm wide treated softwood boards

Bituminous adhesive

Cement and sharp sand screed

Fig. 11.8 Hardwood flooring to a sports hall

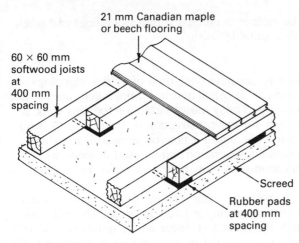

Fig. 11.9 Hardwood floor construction for squash courts

Special effect concrete

The appearance of concrete floors may be improved by coloured cements and variations in the aggregate content. Standard concrete mixes can be improved by abrasive surface treatment to expose the aggregates. This may be achieved by carborundum attachments to a power float.

Granolithic concrete

Granolithic concrete is a mixture of cement, fine aggregate and granite or whinstone chippings in the ratio of 1 : 1 : 2. It is most suited to areas of heavy pedestrian traffic and industrial use, as the concrete has excellent resistance to chipping, cracking and abrasion. It is applied as a finish to matured or freshly laid concrete. If the former, the base concrete surface must be thoroughly cleaned, keyed and grouted with a cement slurry to encourage bonding. The granolithic topping should be laid to at least 40 mm thickness. Monolithic construction is preferred with the base concrete still fresh as only a 20 mm topping is necessary. Bay sizes will depend on the position of construction joints in the sub-base. If the base is at least 150 mm thick, each bay should not exceed 30 m²; if only 100 mm thick, the bay size should not exceed 15 m². Joints between bays of granolithic finish need no special treatment and may be plain butt joints, but they must coincide with any sub-base joint. Finishing is by trowelling at least three times at 2- to 3- hour intervals to produce a hard, uniform surface with excellent durability and abrasion resistance.

Sprinkling the surface of wet concrete with a dry mixture of cement and granite (1 : 2) produces a similar finish, but the wearing layer is thinner. The surface is tamped with a wood float and steel-trowelled to achieve a suitable finish.

Terrazzo

Terrazzo is an attractive concrete composed of marble chippings and white or coloured cement. It originated in Venice where marble and mosaic work is abundant. Today most marble for terrazzo work is still exported from Italy for the construction of feature floors, walls, stairs, etc. The construction and finishing of terrazzo is a highly skilled and time-consuming operation, rarely used in modern building as its application is incompatible with rapid construction techniques.

The aggregate is crushed marble; it should be angular and not less than 3 mm, otherwise crazing is likely to occur. The mix ratio should be 1 : 2½ to 3; the slightly stronger mix is suitable for smaller aggregates. Aggregates are graded in the size ranges shown in table 11.2

Table 11.2

Grade	Size (mm)
00	1½–2½ only suitable in very small panels
0	2½–3
1	3–6
2	6–10
3	10–12
4	12–20
5	20–25

Terrazzo is applied to a sub-base of normal concrete, preferably monolithically, i.e. within 3 hours of casting the base. A 12 mm layer of terrazzo is formed over the concrete which is divided into panels of 1 m² maximum area. The panels are formed by brass dividing strips to prevent the terrazzo cracking. Alternatively, the concrete sub-base is allowed to harden, the dividing strips are placed and levelled, and a 15 mm screed of 1 : 3 cement and sharp sand is laid over the base. The terrazzo topping is placed while the screed is still wet. It is tamped to ensure consolidation and lightly trowelled to provide a flat surface. After careful curing, and the surface has hardened sufficiently to prevent damage, it is ground with coarse carborundum and water followed by finer grinding and polishing at 3- to 4-day intervals. The detail in Fig. 11.10 shows the laying procedure on a dry concrete sub-base.

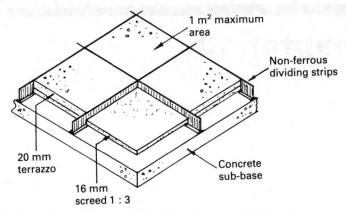

Fig. 11.10 Monolithic terrazzo floor construction

Chapter 12

Specialised paints and wallcovering materials

Paint

The composition and application of paints are considered in Chapter 14 of *Construction Processes, Level I*. This section considers some specific function paint finishes.

Anti-condensation paint. This contains fillers of vermiculite or cork dust. It provides some thermal insulation and functions as a moisture absorber. The effectiveness is limited to saturation of the filler and these paints must not be considered as an alternative to adequate ventilation.

Anti-glare paint. This is designed to reduce solar glare and heat transfer through glazing and roof lights. It is water-based for easy removal by brushing after the summer. Common applications include glazed factory roofs and greenhouses.

Bituminous paints. There are produced from bitumen, asphalt or petroleum residues, dissolved in white spirit or paraffin hydrocarbons. These solvents are strong and where bitumenous paints are applied over oil-based paints they must be sealed to prevent discolouration. Similarly, oil paint should not be applied over bituminous paint as it remains semi-liquid for a long time. If exposed to sunlight, hardening and surface cracking will occur unless protected with a bituminous aluminium paint surface. Bituminous paints are very useful for protecting steelwork from corrosion, particularly when submerged in water or below ground.

Cement paint. This is available in powdered form for mixing with water just before brush, trowel or spray application. The material is basically white cement with colouring pigments and other additives to improve mixing and water-proofing. These paints are a suitable water-repellent treatment to rendered and masonry walls.

Flame retardant paint. This is produced to upgrade combustible surfaces by retarding the ignition time and reducing the spread of flame effect. Class O and Class 1 (BS 476 : Part 7, and Building Regulations; part E) standards are possible on softwood with three- and two-coat applications of fire-resistant and intumescent paint. Fire-resistant paints of water or oil composition are not easily flammable and can be made more flame retardant by inclusion of antimony oxide and chlorinated or brominated compounds. Alternatively, intumescent varnishes or emulsions may be used. Varnishes have the advantage of retaining the authenticity of exposed timber beams and similar features. These paints swell on contact with flames and produce a charred but stable carbonaceous foam which thermally insulates the combustible substrate from the flame.

Floor paints. These function as a sealant for concrete floors and also provide a tough and abrasion-resistant finish. They are available in one- and two-pack polyurethanes or epoxy resins. Two-pack paint systems require a catalyst mixed with the paint. These types are less convenient than the one-pot system, but have greater cohesion and better wearing properties.

Fungus-resisting paints. These are used where mould growths could germinate. Contemporary construction techniques should prevent the possibility of dampness and condensation penetrating the building fabric. However, some structures, notably swimming pools and breweries, will suffer the effects of condensation and moisture penetration.

Gloss and emulsion paints. These may incorporate fungicides, and for particularly damp areas where heating and ventilation cannot overcome the problem, specially formulated paints are produced. Before reconstruction of existing contaminated areas all fungal growths and suspect timber must be removed. Masonry and concrete should be heat treated with a blow lamp and liberally brushed with a sodium fungicide solution (sodium pentachlorophenate, sodium orthophenylphenate or sodium hypochlorite).Household bleach is a less effective but often adequate substitute.

Insecticidal paints. These have been developed to kill insects on contact and by digestion. They are available as decorative varnishes and coloured paints. Where prevention of woodworm infestation of timber is required (Building Regulations, Part B) vacuum-pressure impregnation of structural timber should precede delivery to site. These insecticide treatments are not colourless and they may contain harmful chemicals, hence their unsuitability on timber exposed to habitable areas.

Multi-colour paints. These produce an irregular surface containing a fleck or spatter finish. The flecks are globules of colour having a different binder than the main paint component. Application of gloss and emulsion paints is most successful by spray. The result is a hardwearing finish used in schools and similar situations where exclusion of dirty marks and surface deformities is necessary.

Road paints or line-marking paints. These are based on chlorinated rubber or one-pack epoxy resins. They offer excellent resistance to abrasion and chemical attack. An alternative is hot application of plastic or bitumen-based solutions as defined in BS 4147. Line paints may contain fine glass granules to produce a reflective effect from vehicle headlights.

Textured paints. These are heavily bodied paints produced by mixing soluble resins with water until a 'plastic' consistency is obtained. They are also available as ready-mixed emulsions. Thick applications are made with brush or trowel to ceilings, and finishing is by texturing with a sponge, brush or other implements to provide an irregular or patterned effect. These paints are useful for filling and concealing surface defects.

Wallcoverings

The choice of wall decorating material is extensive, including traditional wallpapers, foils, textiles and plastic sheet materials.

Wallpapers
Lining paper. This is used to provide uniform suction on uneven surfaces and to reinforce cracked plaster before applying the decorative wallpaper finish. The quality varies from common pulp for normal use to a stronger brown paper for badly cracked surfaces, and pitch-impregnated paper for damp walls. Pitch-lining paper must not be regarded as a cure for dampness and it is not intended as a substitute for remedial treatment to a source of dampness. A cotton-backed paper is also available for boarded backgrounds likely to move. Lining papers are produced in rolls 11 m long × 560 or 760 mm in width. They should be hung square to the finishing paper

to avoid coincidence of joints. This is known as 'cross-lining'.

Decorative finish wallpapers vary considerably in quality, composition and price. They are produced rolled in 10 m lengths × 530 mm wide. Patterned papers repeat; this refers to the distance between identical patterns which may be as much as 2 m. This considerably affects the quantity of paper used as a large repeat will generate more wastage than a small repeat or non-patterned paper. Some popular examples of wallpaper are listed below.

Flock papers. These are manufactured with a raised pile pattern produced by blowing coloured wool, cotton or nylon fibres onto an adhesive printed pattern.

Metallic finishes. Paper-coated with metal powders or a metal foil sheet material.

Moires. A watered silk effect finish.

Ingrained. An effect produced by incorporating sawdust or other fibres in the paper surface.

Textured effect. An imitation texture achieved by embossing the surface.

Washable papers. These include varnish and plastic-coated paper, in addition to expanded PVC foam and textured PVC sheet.

Relief papers. These are divided into the two categories of high relief (anaglypta) and low relief (lincrusta). Anaglypta is a quality wallpaper moulded whilst wet to provide embossed patterns which give the paper strength and resilience. The depressions on the back of the paper should be filled with paste and sawdust to resist impact damage. Lincrusta is a paper composed of whiting, wood flour, lithopone, wax, resin and linseed oil which is superimposed hot, on kraft paper. The result is relatively dense paper requiring a heavy-duty paste to successfully adhere it to the wall.

Textiles

These are a very fashionable wallcovering produced in hessian, linen, silk, synthetic and natural fibres. Widths vary depending on the manufacturer, and rolled lengths may exceed 50 m. Colours are natural or dyed and appearance can be varied by closeness of weave and blending of strand sizes. Woven textiles are available unbacked for direct adhesion to the wall or backed in similarity to stranded textiles, with paper or foamed plastic.

Felt and simulated suede are alternative attractive wall coverings. They are readily dyed to provide a smooth, richly coloured luxury finish.

Expanded polystyrene is available in rolled sheets of 610 mm width × 2 to 5 mm thickness. It functions as a lining paper with the advantage of some thermal insulation. Alternatively, it may be used as a finish, decorated with emulsion paint. Tiles are also produced, generally for application to ceilings. Both tile and sheet effectively hide surface defects, but behave badly in fire. The material flames readily and drops of burning plastic spread the fire to other parts of the building. Fixing is with a PVA adhesive only, as other glues may dissolve the polystyrene.

Wall preparation

Before commencing wall covering the surface should be dry, sound and clean. Cracks and defects should be filled with a plaster filler and sanded smooth. New buildings must be given several weeks to dry out. The extent of time will depend on the time of year and the period which the building has been vacant. Existing paper is removed by wetting and scraping with a stripper. Pretreatment may include application of fungicide and size. Fungicide or mould inhibitor is diluted in water and brushed over the wall surface. Size is similarly applied to seal the plaster and reduce the suction effect. Modern pastes contain both fungicide and size, but the amount is limited. Pastes are products of flour, starch or cellulose and are normally applied to the back of the wallpaper prior to hanging. Plastic lining materials are usually applied to pasted walls. Paste staining of the paper surface should be avoided particularly on flock papers and textiles. Irregular application of paste may also cause surface staining, uneven adhesion and damage to textile finishes. With textiles, the manufacturers recommended adhesive should be employed as the backing material may be unsuitable for normal pastes.

Chapter 13

Domestic spatial requirements

The internal layout and arrangement of facilities within a building should be determined by the dimensional characteristics of the human body and the number of people using the facilities. It should also incorporate versatility, to adapt to changing living patterns and variable family requirements. The study of anthropometry provides data relating to body dimensions and reach characteristics. Some of the most significant average dimensions which affect the design of equipment for human use (ergonomics) and its layout are shown in Fig. 13.1.

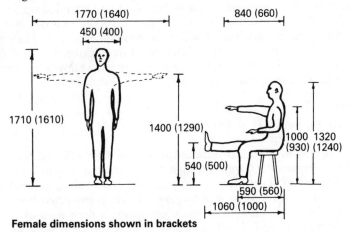

Female dimensions shown in brackets

Fig. 13.1 Average human dimensions

Space in the home

Following the recommendations of the government publication
Homes for Today and Tomorrow (Parker Morris Report, HMSO
1961) there has been a new approach to provide improved living
standards relating to the family size. Much of this has been
incorporated into the spatial layout of habitable rooms. The
dimensions indicated in this chapter are provided as guidance only,
as it is impossible to precisely adapt variable human space
requirements to the diversity of modern constructional design.

Bathroom

The minimum provision of WCs depends on the number of people
residing in a dwelling. Table 13.1 indicates the requirements.

Table 13.1

Persons	WC requirements
Up to 3	1 – may be in bathroom
4 or 5	1 – must be separated from bathroom
6	2 – 1 may be in bathroom

Furthermore, where a WC is not contained within a bathroom it
must have wash-basin facilities.

The design of bathrooms is not necessarily related to a minimum
floor area but the minimum activity space for use of appliances.
However, dressing and drying space must be incorporated into the
design of bathrooms containing a bath or shower and this will
occupy an area of up to 900 × 1100 mm, which may overlap activity
space. The space occupied by appliances will vary a little between
manufacturers. For purposes of illustration the sizes shown are
extracted from the appropriate British Standard and rounded up to
the nearest 100 mm. Overlap of activity space is permitted as it is
unlikely that adjacent appliances will be in simultaneous use. The
amount of activity space shown is determined from extreme
dimensions required in the various postures and is to allow for clean
and comfortable use of each appliance.

Wash-basin

Overall size on plan 600 × 400 mm.
Height to rim, 800 mm.
Rim heights may vary to suit different uses. Children will find the
pedestal set height of 800 mm too high and graduated heights will be
useful in schools to suit different age groups. In the home, access for
small children is achieved by standing on a firmly constructed box.
Figure 13.2 illustrates an activity space occupying a floor area of
1000 × 700 mm to the full ceiling height. This is sufficient for both

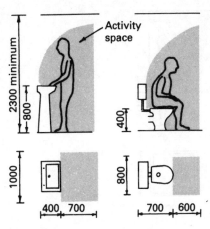

Fig. 13.2 Wash basin and WC

hand and hair washing provided the area immediately above the basin is clear of obstructions.

Water closet

Overall size on plan, including the flushing cistern, 500 × 700 mm. Height to rim, 400 mm.

The height of the pan is a compromise between the requirements for satisfactory seated use and male standing use. Activity space is shown in Fig. 13.2 measured to the full ceiling height over an area 800 × 600 mm, which is also the recommended space for a bidet installation.

Bath

Plan size, 1700 mm long × 700 mm wide;
Height to rim, 500 to 600 mm.
The amount of activity space adjacent to a bath should be sufficient to permit comfortable access and egress. It should also allow sufficient space for drying in addition to providing enough space for an adult to attend a child. Clear floor space occupying 1100 mm along the bath and 700 mm to the side, preferably at the tap end, is sufficient extending to the full ceiling height. The remaining space shown with the activity space in Fig. 13.3 could accommodate a cupboard or laundry basket.

Shower

Plan size, 900 mm × 900 mm.
Tray height, 150 mm.
The amount of access and drying space required will extend 700 mm beyond the shower and occupy the full tray width. Where

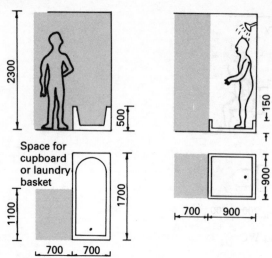

Fig. 13.3 Bath and shower

the shower is not walled at the sides more freedom for movement is possible and the space may be reduced to 400 mm.

In similarity with other sanitary fittings, the full ceiling height is absorbed within the activity space and where a bath or a shower is provided the full dressing space of 1100 × 900 mm will be necessary. Requirements for a shower are shown in Fig. 13.3.

Bathroom layouts

The arrangement of sanitary appliances and possible combinations are numerous. For planning purposes, pieces of card cut to scale to represent the appliance area and activity space could be superimposed on the room plan and moved around until the optimum arrangement is found. Activity space may overlap, unless the room is for multiple use but dressing space should only marginally overlap appliances. A few possibilities for two- and three-appliance layouts are shown in Fig. 13.4 and 13.5 with the overall internal dimensions. The arrangement of sanitary pipework must also be considered particularly with regard to maximum branch lengths and falls. These requirements are explained in Chapter 8.

Kitchen

Kitchen design and layout is now highly professional, particularly with the range of gadgetry and machinery expected in the modern kitchen. Legislation governing minimum standards of space and storage requirements is extensive. Public sector construction is influenced by the requirements of the Ministry of Housing and Local Government circular No. 27/70, *Metrication of housebuilding*

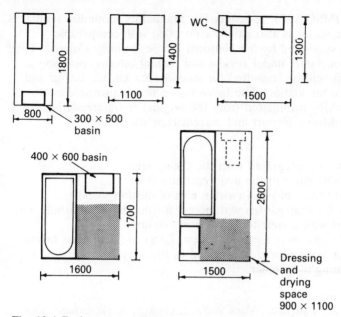

Fig. 13.4 Bathroom and WC layout – two appliances

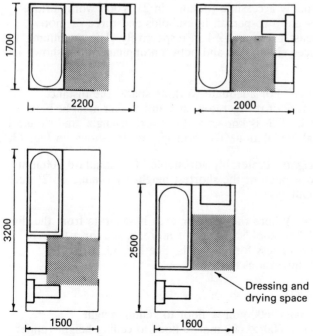

Fig. 13.5 Bathroom and WC layout – three appliances

progress, HMSO 1970, appendix IV, Mandatory minimum standards. The private sector is also strictly controlled with comparable standards established by the National House Building Council. Their rules are constantly under review and current internal planning schedules should be consulted for guidance on kitchen layout and storage accommodation. The following section is provided as a design checklist formulated from the original recommendations of the Parker Morris Report and current legislation.

Location
1. Direct to, or integrated with the dining area.
2. Reasonably close to the main entrance door.
3. Close to refuse disposal (within 6 m of dustbin storage)
4. Next to the rear garden with access for putting out washing and provided with a view for supervising children.
5. Natural light, with a pleasant view. Large areas of glass facing South or West should be avoided as this may encourage overheating in summer.

Layout design
1. Unbroken sequence. Work surface, cooker, work surface, sink, work surface: this should be continuous with no gaps or obstructions. Areas of food storage and preparation may be separate provided they are reasonably accessible (defined in 2). Total work surface length for one- and two-person households and three or more persons is at least 1.9 m and 2.1 m respectively. The minimum acceptable spaces either side and between appliances are shown in Fig. 13.6.

2. Working space. For adequate working space, the distance between food store or refrigerator, sink and cooker must be between 3.6 m and 6.6 m. This is known as the 'work triangle' and should be clear of through circulation. The principle is also shown in Fig. 13.6.

3. Worktop heights. Preferably adjustable. The standard 900 mm applies only to women of the shortest height who make up 2½ per cent of the population.

4. Informal use. Where the working area is separate from the dining area, allow, in households of two, or more, space for a table or built-in counter for less formal meals. See Fig. 13.7 for recommended dimensions.

5. Appliance space.

(a) Cooker – 600 × 600 mm × floor to ceiling height.
(b) Refrigerator – 600 × 600 mm × floor to ceiling height (under worktop for one and two-person dwellings).

Total worktop = 1.9 m minimum for one and two person households.
= 2.1 m minimum for three or more persons.

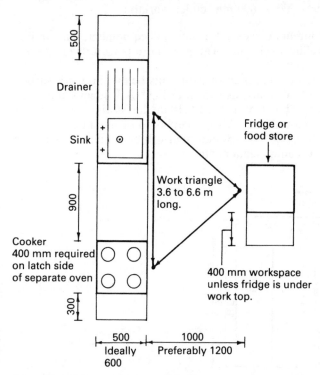

Fig. 13.6 Minimum worktop dimensions

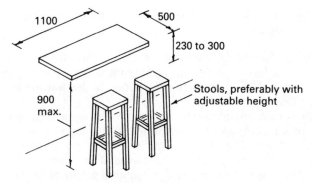

Fig. 13.7 Space for casual meals taken in the kitchen

(c) *Laundry equipment* – one- and two-person dwellings, 900 × 600 mm under worktop.
Over three persons, 1400 × 600 mm × floor to ceiling height.
(d) *Dishwasher* – 600 × 600 mm under worktop.

6. *Storage.* Minimum enclosed for food and equipment: one- and two-person dwellings – 1.7 m³. Three or more person dwellings – 2.3 m³.

Storage may be composed of base units below the work surface, wall cabinets and tall units, as shown in Fig. 13.8. Ideally this space should include a cabinet of at least 0.18 m³ for storing vegetables at a temperature below 12 °C. Also, a cupboard of minimum dimensions, 500 × 600 × 1500 mm high will be required for brooms, ironing board, vacuum cleaner, etc.

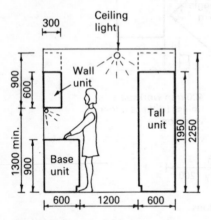

Fig. 13.8 Standard kitchen units

7. *Clearances.* See Fig. 13.9 for a description of activity space and safety regarding clearance.

8. *Artificial lighting.* Should provide 200 lumens/m² (lux) to illuminate working surfaces. Ceiling fittings are only really suitable as background lighting with working lights positioned under wall units as shown in Fig. 13.8.

Natural light is best provided from a window above the sink unit as the majority of kitchen tasks are performed in this area or on associated work surfaces. The window area is limited by wall space requirements and maximum acceptable heat loss. Where windows are provided a clear zone of open space is required adjacent to the external face of the wall. The Building Regulation requirements for area of glazing and open space are detailed in Appendix to this chapter.

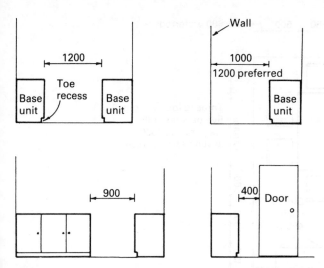

Fig. 13.9 Recommended clearances around kitchen units

9. Ventilation. Where windows are provided an openable area equal to at least one twentieth of the floor area is required. Mechanical extraction by fan is an alternative, which requires a fresh air inlet with free opening of at least 0.15 m² to replace extracted air. This should be strategically placed to encourage a crossflow of air. Fans may also be provided in ducted hoods over the cooker. These are the most efficient means for disposing of steam, fumes and excess heat. Ventilation to larders is provided by high and low level air bricks, permitting at least 4500 m² of air to circulate through.

10. Services. Sufficient space is required for a radiator to maintain a temperature of 13 °C in working kitchens. Other piped services to appliances are more easily located to 'in line' arrangements than diverse layouts. Excessive hot water 'dead-legs' (see Ch. 6) must be avoided in addition to long and steep waste pipes (see Ch. 8). At least four electric power sockets are necessary and where the combined loading is high, particularly from tumble driers and washing machines, a separate ring main or spur is advisable.

Living areas
Space requirements in living areas are impossible to define and less easy to calculate than bathrooms and kitchens as the dimensions of fittings and furniture differ considerably. The illustrations in Figs. 13.10, 13.11, 13.12 and 13.13 are intended for guidance only and indicate approximate requirements for furniture and adequate clearance space between items of furniture and perimeter wall.

This subject is extensive as well as diverse, and therefore only

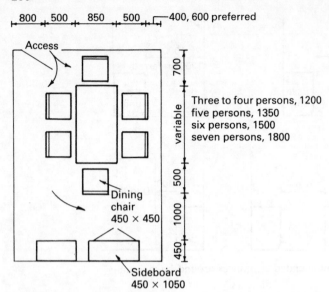

800 | 500 | 850 | 500 | 400, 600 preferred

Access

700

variable

Three to four persons, 1200
five persons, 1350
six persons, 1500
seven persons, 1800

500

Dining
chair
450 × 450

1000

450

Sideboard
450 × 1050

Note. Dimensions shown are adequate and variable, overall
dimensions should dimensionally co-ordinate with structural
units.

Fig. 13.10 Dining-room accommodation for six persons

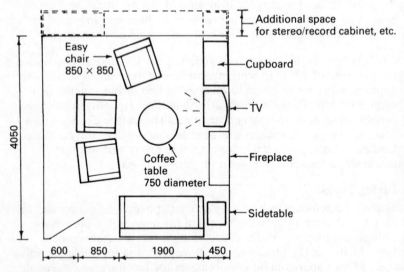

Additional space
for stereo/record cabinet, etc.

Easy
chair
850 × 850

Cupboard

TV

4050

Coffee
table
750 diameter

Fireplace

Sidetable

600 | 850 | 1900 | 450

Fig. 13.11 Sitting-room accommodation

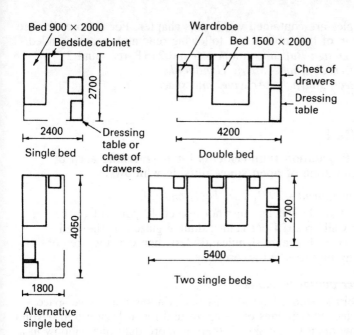

Fig. 13.12 Bedroom accommodation, allowing space for bed making and dressing

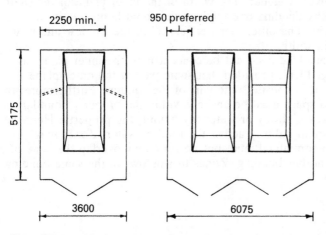

Note. 90 per cent of cars are not longer than 4.8 m, nor wider than 1.85 m.

Fig. 13.13 Garage space

the principles are contained within this chapter. For a more detailed appreciation of the subject, the following reading is recommended:

DOE Design Bulletin, numbers 6 and 24 (Parts 1 and 2).

Architects Journal – Metric Handbook.

Architects Journal – Activities and space.

Appendix I

Building Regulation requirements for permissible area of glazing and zone of open space outside windows.

Second amendment – 1981, Part F6(1) and (2).

The area of single glazing must not exceed 12 per cent of the perimeter wall area, 24 per cent if double glazed. If the wall structure has a 'U' value significantly less than 0.6, the area of glazing may be increased accordingly.

Open space outside windows

Every habitable room with at least one window must have space unobstructed by buildings or rising ground for at least 3.6 m beyond the window open to the sky. If there is more than one window in a room, the zone of open space may be shared. This imaginary shaft is illustrated in Fig. 13.14. It rises at an angle of 30 ° from the lower window level, i.e. the lowest level of glazing or 1.2 m above floor level, whichever is higher. The width of the inner plane of the shaft is calculated by dividing one tenth of the room floor area by the window height. The outer plane is parallel to the window and is at least 3.6 m in width.

The outer plane need not be concentric to the inner plane, as shown in Fig. 13.15, provided that some part of the outer plane remains directly opposite some part of the window. Furthermore, the zone of open space must be entirely within the property boundary except where a highway or waterway bounds the property. Here the zone may extend halfway across the road or water. Zones may overlap onto communal land, but no zone may overlap the zone of a window in another building. Zones to windows in the same building may overlap.

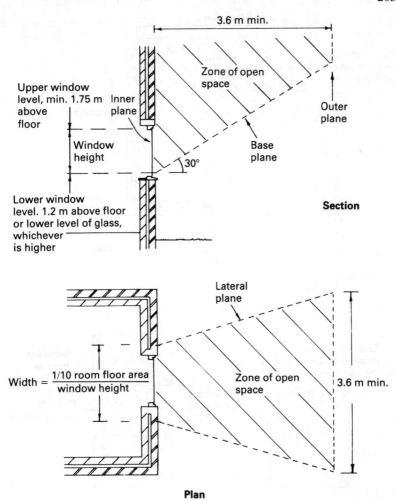

Fig. 13.14 Shaft of open space required outside a window to a habitable room

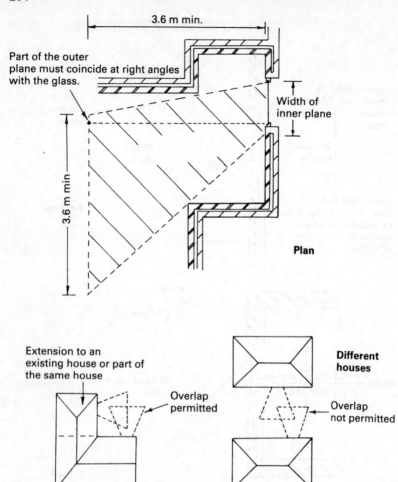

Fig. 13.15 Application of the space adjacent to a window

Chapter 14

Location and fixing techniques

Fixing devices

Nails

These are detailed in BS 1202 : Pts 1, 2 and 3 : 1974, covering steel, copper and aluminium respectively. Surface and head treatment varies as shown in Fig. 14.1 to suit different applications. Most examples are manufactured in galvanised steel for use in corrosive circumstances where copper or aluminium would have inadequate strength, e.g. roof timbers treated with insecticide. When ordering, the following should be specified:

Material.
Amount, in kilograms.
Type – see Fig. 14.1.
Finish – bright, galvanised or other coating.
Length.
Diameter.
BS number.

Screws

Wood screws are manufactured from cold drawn wire of steel, brass, stainless steel, aluminium, silicon bronze and nickel-copper alloy to BS 1210 : 1963. Numerous finishes are possible for protective and decorative purposes. These include non-ferrous plating, polishing and laquering.

Standard screws are produced in 24 lengths ranging from 3 mm

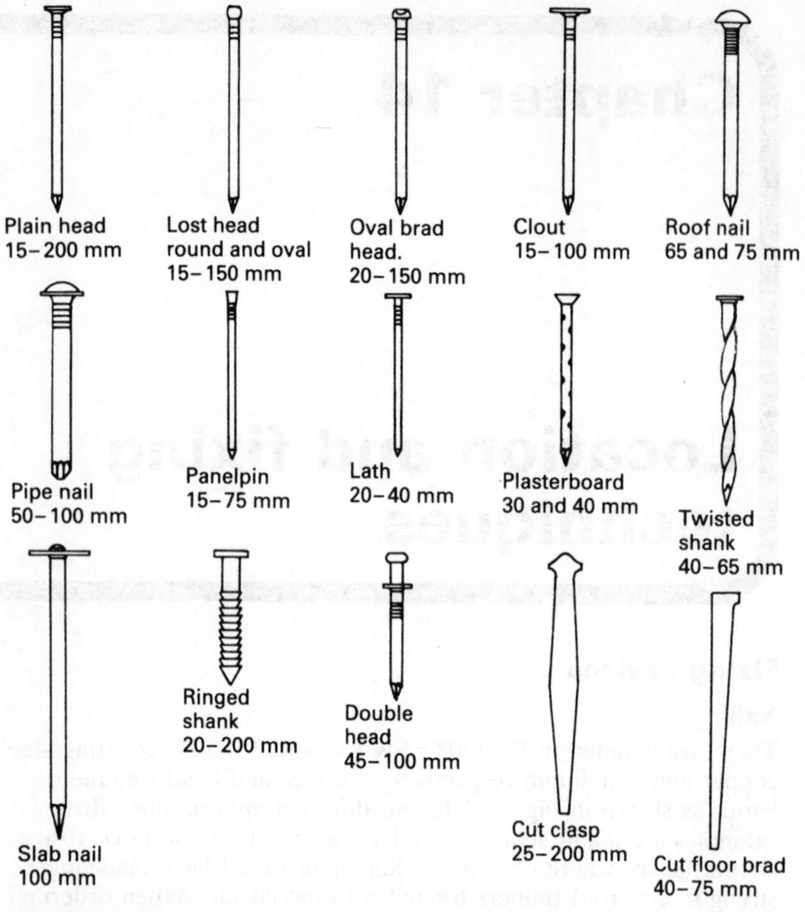

Fig. 14.1 Nail types

Plain head
15–200 mm

Lost head
round and oval
15–150 mm

Oval brad
head.
20–150 mm

Clout
15–100 mm

Roof nail
65 and 75 mm

Pipe nail
50–100 mm

Panelpin
15–75 mm

Lath
20–40 mm

Plasterboard
30 and 40 mm

Twisted
shank
40–65 mm

Ringed
shank
20–200 mm

Double
head
45–100 mm

Cut clasp
25–200 mm

Cut floor brad
40–75 mm

Slab nail
100 mm

up to 152 mm and in 19 diameters from 1.52 mm to 12.7 mm. Screw diameter is expressed as screw gauge, the smallest being 0 increasing in consecutive numbers up to 10, and then alternate numbers up to 20. The three largest diameters, 24, 28 and 32, are very rarely specified. Gauge and length are distributed so that the smallest diameter screws are available in the lowest gauges and the longest screws in the higher gauges.

The following should be specified when ordering:
Material
Quantity (No.)
Head (see Fig. 14.2)
Length
Gauge
BS number
Finish or plating

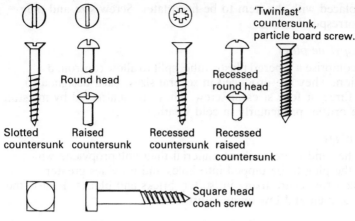

Fig. 14.2 Screw types

Securing

In hardwoods and dense particle boards, pilot holes of up to about 90 per cent of the screw diameter should be provided. In softwoods

Table 14.1

Gauge	Diameter	Gauge	Diameter
0	1.52	8	4.17
1	1.78	9	4.52
2	2.08	10	4.88
3	2.39	12	5.59
4	2.74	14	6.30
5	3.10	16	7.01
6	3.45	18	7.72
7	3.81	20	8.43

and less dense particle boards, holes of about 65 per cent screw diameter are adequate. Table 14.1 compares the screw gauge with the nominal diameter in millimetres.

In hardwoods, pilot holes are unnecessary for screws below gauge 3 and in softwoods below gauge 6.

Light wall plugs (see Fig. 14.3)

These are inserted into drilled holes in masonry and concrete to permit the transmission of load from wood screws or coach screws.

Fibre-plugs

These are jute fibrous tube impregnated with a waterproof binding agent, treated against bacteria and fungi. The screw forms its own thread, expanding the plug without damage and may be withdrawn and replaced with the item to be fixed later. Screw size and plug must correspond.

Polypropylene plugs

These comprise a tapered plastic tube, split to allow controlled expansion. They are produced in several sizes, each designed to accept three or four sizes of screw. They are unaffected by moisture, but are brittle, particularly in cold weather.

Nylon plugs

A tougher and more resilient material than polypropylene which allows the plug to be tapped into holes and provides greater expansion for secure fixing in 'softer' bricks and blocks. Resists the effects of high and low temperature.

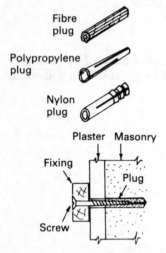

Fig. 14.3 Screw fixings to solid backgrounds

Asbestos fibre compound

This material is mixed with water to produce a plastic material for insertion into rough or irregular shaped holes. The screw may be inserted immediately. It has a limited bearing capacity.

Cavity fixings (see Fig. 14.4)

These are necessary where it is impractical to locate fixtures to the timber framework in ceilings and partitions, and also for timber-framed exterior walls, where access to only one side of the fixing material is possible.

Expanded rubber sleeve

This is composed of a rubber bush incorporating a nut which is drawn towards the back of the wall lining by rotation of the screw. The sleeve compresses to form a sound expanded fixing which permits the screw to be withdrawn and reinserted if required. They may also be used as a fixing to cellular concrete blocks. In all situations they have the advantage of non-corrosive, waterproof, electrically insulative and vibration-proof qualities.

Spring toggles

These contain two toggle bars hinged and sprung about a swivel nut located on a screw. The toggles are pushed through a hole in the wall lining where they spring open and, after tightening the screw, provide a wide bearing on the inside. The toggles are lost if the screw is removed.

Gravity toggles

A toggle bar is hinged about a swivel nut on a screw. One end of the toggle is heavier than the other, so that when the toggle is placed through a fixing hole it hangs vertically. Tightening the screw pulls the toggle back against the inside of the lining. Screw removal will lose the toggle.

Umbrella anchors

These steel anchors contain a threaded nut connected by legs to a flange. The device is placed through the fixing hole until the flange bears against the lining. Tightening the screw causes the legs to fold at predetermined points until they bear on the back of the lining. The screw may be removed as these are designed to provide permanent fixings.

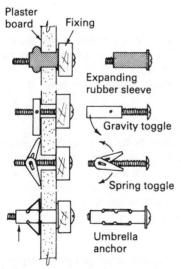

Fig. 14.4 Hollow fixings

Expanding bolts

These are manufactured in a variety of forms, each relying on the frictional grip between an expanding cylindrical shell and surrounding masonry or concrete. The device, available with a projecting thread or for bolt insertion into an expander, is placed within a hole as shown in Fig. 14.5. The fixture is positioned and the bolt or nut located and tightened until the expanding shell obtains sufficient grip. This is primarily intended as a heavy-duty masonry fixing. Also shown in Fig. 14.5 is a nail-in expanding fixture. This is an ideal vandal- and thief-proof fixing as it is very difficult to remove.

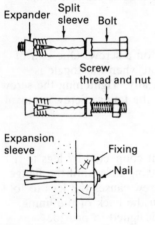

Fig. 14.5 Expanding fixtures to masonry

Built-in fixings

These are either bolts or sockets cast into the structure. They must be accurately located by a jig or template or positioned in formwork before the concrete is placed as shown in Fig. 14.6. For lighter fixings with screws or nails the composition blocks also shown in Fig. 14.6 may be used.

Resin anchors

These are a method of chemically bonding bolts to masonry. A glass capsule containing polyester resin, quartz granules and a phial of hardener is placed inside a drilled hole. A threaded stud is attached to a drill and inserted in the hole to break the capsule and phial. The drill is rotated until the resin and catalyst are mixed and the stud is left to bond to the structure as shown in Fig. 14.7.

Cartridge guns

These are gaining in popularity as an alternative means of securing

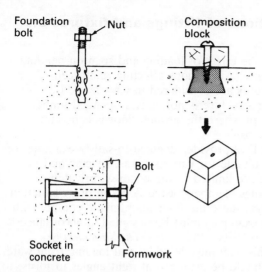

Fig. 14.6 Cast-in fixtures

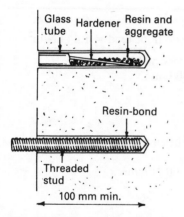

Fig. 14.7 Chemical bonded stud

fixtures. They are a very quick and efficient method most suited to structural fixings for timber, steel and concrete. Steel sections may be penetrated to a depth of 25 mm and the variety of studs, clips and fixings is considerable. Manufacturers provide guidance in the use of this equipment and untrained operatives should not use these tools.

Structural implications of fittings and fixtures on timber floors

Simple interpretation of the effect of loading and span on bending moment and other design considerations affecting reinforced concrete and steel structures are considered in Chapter 3. In dwelling houses the floor structure is normally suspended timber joists and boarding, except where the ground floor is concrete, bearing directly on the subsoil.

Building Regulation D12 provides deemed-to-satisfy examples of structural floor timbers suitable for various spans, spacing and loading. However, these are limited to three-storey, privately occupied domestic buildings. Timber-joisted floors in other situations or those used in dwellings where the floor load fails to comply with the Building Regulation examples must be designed in accordance with the requirements of BS 5268 : Pt 2 : 1984, *The Structural Use of Timber*. In housing this may include additional loading from water tanks and baths which should be arranged at right angles to joists, to distribute the load over several units.

Example of timber floor design
Building Regulations, maximum imposed floor load = 1.44 kN/m^2

Data: Loading = 2 kN/m^2 (i.e. in excess of Bldg Reg.)
Span = 5 m. Joist spacing = 400 mm.
Timber = ss grade Hem-fir.

From BS 5268:
fibre stress f = 7.3 N/mm^2
compression perpendicular to the grain = 1.55 N/mm^2
shear v = 0.8 N/mm^2
modulus of elasticity, E = 10 700 N/mm^2.

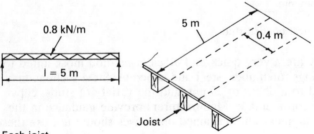

Each joist
supports an area 5 m × 0.4 m = 2m^2
Superficial load = 2 kN/m^2
therefore joist load = 2 × 2 = 4 kN
= 0.8 kN/m span.

Fig. 14.8 Timber joist design

A suitable joist must be designed for: (a) resistance to bending; (b) deflection; (c) shear.

(a) The bending moment formula for a simply-supported uniformly loaded beam is Wl/8 (see Fig. 3.33), and the design formula is equal to $f \times Z$ where Z, the section modulus $= bd^2/6$

Thus   BM $= Wl/8 = fbd^2/6$

where   BM $=$ bending moment
           W $=$ load
           f $=$ fibre stress
           b $=$ breadth of section
           d $=$ depth of section

           W $= 2 \times 10^3 \times 5 \times 0.4 = 4000$ N
           l $= 5 \times 10^3 = 5000$ mm

Transposing the formula: $\dfrac{Wl}{8} = \dfrac{fbd^2}{6}$

$bd^2 \qquad = \dfrac{6Wl}{8f} \qquad = \dfrac{6 \times 40000 \times 5000}{8 \times 7.3}$

$\qquad\qquad = 205\ 4794$ mm^3

If b $= 50$ mm

then d $= \sqrt{\dfrac{205\ 4794}{50}} \quad = \begin{array}{l} 203 \text{ mm} \\ \text{nearest size is } \underline{225 \text{ mm}} \end{array}$

(b) *Deflection.* The maximum allowed by BS 5268 is 0.003 of the span. In this example 0.003×5 m $= \underline{15\ mm}$. Check, using 50×225 mm joist.

For uniformly distributed load the formula is

$\dfrac{5}{384} \quad \dfrac{Wl^3}{EI}$ (see Fig. 3.35)

$\qquad\qquad\qquad I = \dfrac{bd^3}{12}$

$\qquad\qquad\qquad = \dfrac{50 \times 225^3}{12}$

$\qquad\qquad\qquad = 4.7 \times 10^7$ mm

Deflection $= \dfrac{5 \times 4000 \times 5000^3}{384 \times 10\ 700 \times 4.7 \times 10^7}$

$\qquad\qquad\qquad = 12.9$ mm, therefore satisfactory.

End-bearing of joists should be not less than 90 mm for direct bearing on masonry and 75 mm on a wall plate. It is worth checking that these are sufficient.

$$\text{Safe bearing} = \frac{\text{load at the end of joist } (\frac{W}{2})}{\text{compression perpendicular to grain} \times \text{breadth}}$$

$$= \frac{2000}{1.55 \times 50} = \frac{26 \text{ mm}, \text{ therefore } 75 \text{ mm}}{\text{and } 90 \text{ mm are adequate.}}$$

Check for shear strength

$$\text{formula, V} = \frac{2bdv}{3}$$

where
$V =$ vertical load at joist end
$v =$ shear strength of section

thus
$$bd = \frac{3V}{2v} = \frac{3 \times 2000}{2 \times 0.8} = 3750 \text{ mm}^2 \text{ minimum}$$

$$bd = 225 \times 50 = 11\ 250 \text{ mm}^2$$

More than satisfactory.

Holing and notching joists

Fittings and fixtures requiring provision for services inevitably disrupt the structure and possibly de-grade the design strength of certain elements. The area most abused is service access through timber joists. Figure 14.9 shows the location of the neutral axis in a timber section, occurring midway between the maximum compressive and tensile forces. This is the optimum place for holing a joist for electrical cable access but it is impractical to expect pipes to be positioned here, hence pipe location by notching to the upper part of a section. Any form or location of notches will weaken the section and effectively reduce its bearing capacity. However, the extent of damage is reduced if notches are sensibly located to avoid the areas of maximum bending at the centre and maximum shear at each end. Figure 14.9 shows where notches are least damaging, relative to span and maximum depth.

Suspended ceilings

Commercial buildings, hospitals and schools are unlikely to have sufficient provision for absorbing services within the floor structure. In these situations, pipes, ducting, etc. suspend from brackets bolted to the concrete ceiling and are hidden by a false ceiling similarly suspended. Examples are shown in Fig. 14.10.

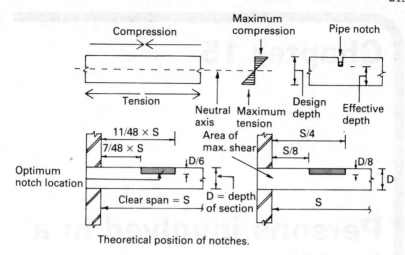

Fig. 14.9 Location of notches in joists

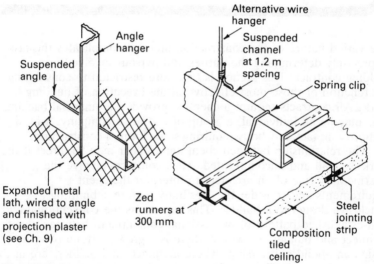

Fig. 14.10 Suspended ceilings

Chapter 15

Persons involved in a building contract

The varied nature of the construction industry complicates the task of precisely determining the parties and persons involved in a building contract. This section is therefore restricted to considering the function of those most effective in the execution of building work. All contracts require a client to provide the financial backing. This may be an individual, a group of people, a company, a local authority; in fact, any body requiring some form of construction.

The relationship between client and builder may be direct if the work is small and uncomplicated. Most contracts involve the intermediary role of an architect to interpret the client's requirements and consider the feasibility, before submitting a conclusive design to a builder. This represents the conventional and well-established pattern of professional commitment between client, architect and builder. However, there is a growing trend towards design-and-build partnerships where architects and builders are in the same employment. This simplifies communications and provides a more efficient service to clients.

Roles and responsibilities of the parties involved in a building contract

The possible inter-relationship of people and organisations involved is shown in Fig. 15.1. The system is simplified, as shown in Fig. 15.2, where the nature of construction is a speculative housing

217

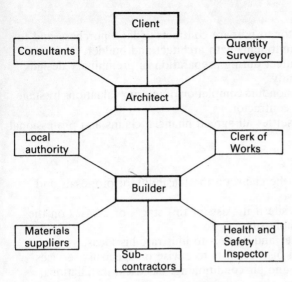

Fig. 15.1 Parties involved with a building contract

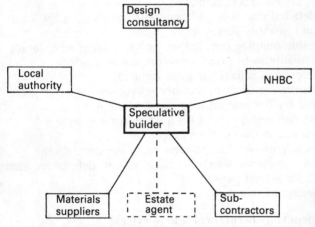

Fig. 15.2 Parties involved with a speculative housing development

development and the builder is the speculator and financial source. Unless the development is extensive or complicated in design, the builder will prepare his own details for local authority approval or possibly employ the services of a design consultancy.

Client

1. Presents the architect with a comprehensive account of his requirements with particular regard to use, space (floor area and volume), cost limit and completion date.

2. Arranges finance.
3. Employs a solicitor to prepare contracts of land purchase and to review legal commitments with architect and builder.
4. Liaises with architect during preparation of preliminary designs and feasibility study.
5. During contract, honours completion of interim valuations by stage payments to the contractor.
6. On completion, settles all agreed financial claims and professional fees.

Architect

1. Initially advises the client on the feasibility of proposals and approximate costs.
2. Investigates the site if it exists at this stage, or advises on the location of possible sites.
3. Prepares sketches and models to illustrate his ideas.
4. Provides a team of specialists to advise on specialist services such as heating and air conditioning, electrical installations, structural requirements, etc.
5. Obtains agreement between client, consultants and himself on size, shape, layout and content.
6. Produces detailed drawings and materials specification for local authority and quantity surveyor.
7. Selects suitable building contractors, provides them with details and bill of quantities to price (work put out to tender).
8. Checks tenders and selects the most suitable.
9. During construction, issues necessary instructions and variation orders to the builder and informs client of progress.
10. Provides interim certificates of completion for the builder to receive payment as the contract proceeds.
11. Certifies completion for full payment from the client to the builder, less a retention which is held to ensure defects are made good within an agreed period.
12. Certifies release of retention.

Consultants (generally structural and services)

1. Advise the designer on the most efficient and economical method of providing their specialism.
2. Prepare detailed drawings within the architect's design.
3. Prepare design calculations and material specifications.
4. During construction, check quality of workmanship and correct use of materials.

Quantity surveyor

1. At pre-contract stage, advises architect on costs and financial viability of certain construction techniques and materials.

2. Prepares approximate costs from architect's initial design sketches.
3. Prepares bill of quantities by measuring and listing the quantities of all materials used in the building in accordance with the current Standard Method of Measurement.
4. Checks builder's priced bill of quantities and advises architect of errors and accuracy of estimates.
5. During construction, prepares stage or period valuations.
6. Agrees the cost of amendments and variation orders with the builder and architect.
7. Prepares the final account.
8. Advises architect of additional costs.

Clerk of works

Generally defined as the architect's representative on the site. His function is to ensure compliance with the architect's drawings, specifications and instructions by checking and inspecting all aspects of construction. The clerk of works reports direct to the architect, as he has no authority over the contractor. He can of course advise and comment on the work to the contractor and assist with difficulties in interpreting the architect's design.

Builder/Contractor

1. Considers whether he has sufficient resources (labour, plant and equipment) and time with regard to other commitments to fulfil the contractual obligations.
2. Conducts a site visit to anticipate difficulties not revealed on the drawings or in the bill of quantities.
3. Obtains and co-ordinates sub-contractor's and supplier's prices and produces an estimate based on these, the drawings, the bill of quantities and the outcome of site investigations.
4. Agrees the estimate with the board of directors or their appointee and submits it to the architect.
5. If accepted, the builder will be given the opportunity to meet with the client, architect and quantity surveyor to agree alterations and amendments.
6. Holds pre-construction meetings to consider allocation of staff and develop contract programme.
7. Co-ordinates programme with architect, sub-contractors and suppliers.
8. During construction, the contractor's roles and responsibilities are considerable. Briefly, they include management of direct staff, sub-contractors, plant and materials during the contracted period of construction with due regard to the relevant codes of quality, safety and legislation. To fulfil these obligations the builder will appoint a manager, generally known as the agent or

'general' foreman, who will also possess the qualities required to harmonise co-ordination between the numerous different agencies on site.

9. At interim stages, and contract completion, provides the quantity surveyor with sufficient data to complete his valuations and accounts.

10. After completion, undertake remedial work within the contracted period.

Local authority (see Ch. 1, of Construction Processes Level 1)

Applications for constructional proposals are received by the appropriate local authority. They are considered by two departments: planning and building control. The planning department's prime function is to ensure that the proposal is suited to the site and does not breach government development restrictions such as 'green-belt' policy. They must also ensure that the design and materials are aesthetically acceptable and that the proposal will not impose excessively on local facilities. These include roads, drains, shops, schools, etc. Most planning legislation is contained in the Town and Country Planning Acts.

The building control section appoints building control officers to particular areas of jurisdiction. Their function is to ensure that the proposal complies with the Building Regulations, with particular regard to suitability and strength of materials. There are specific stages of construction when the official must be given notice to visit the site, although he may call at any time.

Sub-contractors

These are usually employed to provide a specialised or supplementary service. Where the construction requires specialised installations and techniques not undertaken by the main contractor such as lifts, escalators, air-conditioning, fire prevention, etc., the architect prepares a list of suitable contractors agreeable to the main contractor and asks them to submit tenders for the work. Alternatively, the architect may nominate particular sub-contractors known to have the sufficient resources to undertake the work efficiently.

Labour-only sub-contractors are another possibility, generally employed by the main contractor to supplement his own workforce, possibly for the whole contract or at significant stages.

Suppliers: builder's merchants, timber merchants and manufacturers

These are suppliers of general building materials or specialised materials. If the latter they may be nominated by the architect because of their particular style or quality of product. Supplier's responsibilities include:

1. Preparation of quotations.
2. Ensuring material quality meets architect's specifications.
3. Delivery of materials in accordance with the contract programme.
4. Submission of accounts at regular intervals.
5. Provision of agreeable trade and bulk discounts, plus additional 5 per cent if payment is received within 30 days of invoicing.

National House Building Council

This is a non profit-making private organisation created to form a register of house builders capable of producing high standards of construction. The NHBC has its own rules and regulations which supplement the Building Regulations and these are enforced by a team of inspectors distributed throughout the country. Inspectors may examine members' housing projects at any constructional stage, and will issue a ten-year structural warranty to the purchaser on satisfactory completion of the dwelling. The Council's prime objective is to protect the purchaser from disreputable builders and to provide improved standards of accommodation in modern housing. The builder is also protected from undue harassment and unreasonable demands from dissatisfied clients. This organisation has been particularly useful in filling the gap which would normally be occupied by the clerk of works in larger scale architect-controlled construction. Its influence can be appreciated by the fact that very few mortgagers will finance the purchase of non-NHBC guaranteed new housing.

Lines of responsibility within a contractor's organisation

The administrative structure of a building contractor will depend on the size of organisation and nature of work undertaken. The example shown in Fig. 15.3 represents a medium-sized national firm

Fig. 15.3 Contractors organisation

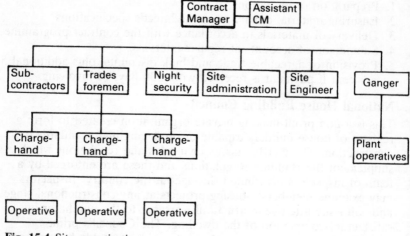

Fig. 15.4 Site organisation

and Fig. 15.4 illustrates the possible site structure with indirect labour provided by sub-contract.

Roles and responsibilities of site personnel

Contract manager

Otherwise known as the site agent because he is the builder's official or contact on site. Some organisations also refer to this person as the site supervisor, controller or general foreman, terms sometimes dictated by the span of responsibility and type and size of contract. He is responsible for all aspects of site construction and personnel administration, duties which are impossible to define, but they could include the following functions:

1. Controlling the contract programme.
2. Organising site administration.
3. Engaging supplementary labour.
4. Chasing and checking material deliveries.
5. Co-ordinating sub-contractors work with his own direct staff.
6. Ensuring health and safety provisions are adequate.
7. Considering clerk of works', safety officials' and inspectors' requests.
8. Organising site meetings with client and architect or their representatives.

Engineer

1. Sets out the initial excavation for foundations.
2. Locates and checks horizontal levels and verticality of the structure.
3. Checks quality of materials, strength of concrete, etc.

Builder's quantity surveyor

1. His responsibilities include: measurement of work completed at specific stages.
2. Valuation of architect's variations.
3. Valuing and paying for sub-contractor's work.
4. Assessment of bonus incentive values.
5. Scheduling and monitoring materials and plant.
6. Processes site cost control information for head office.

Site administration

This can be quite extensive, involving employment of numerous personnel. The timekeeper, storeman and canteen supervisor occur most frequently and their functions are briefly considered.

Time keeper

1. Records attendance of site personnel.
2. Collects cash from bank or co-ordinates collection from a security firm.
3. Prepares wage slips and pay packets.
4. Distributes pay.

Storeman

1. Controls quantities of common materials, e.g. cement, wall ties, etc.
2. Distributes and registers allocation of materials and equipment.

Canteen supervisor

1. Organises weekly menu.
2. Arranges delivery of food and preparation.
3. Prepares financial balance.
Canteen facilities are frequently undertaken by catering contractors to simplify the builder's administration.

Foreman and ganger

Responsible for organising skilled and semi-skilled operatives, respectively. These are rarely entirely administrative and work with their trades operatives. Chargehands are sometimes required to supervise small groups of operatives when the labour content is considerable or widely dispersed.

Operatives

These exist to fulfil the numerous trades in the building industry. Some are better qualified and experienced than others, and to acknowledge this differential slight variations in pay are effected in some of the trades. Assistants, e.g. plumber's mate, exist where craftsmen require semi-skilled help and apprentices are employed from school-leaving age for a period of three to four years' training.

Chapter 16

British Standards, statutory regulations and by-laws

British Standards

The British Standards Institution have published over 1500 documents relevant to the construction industry. Their publications indicate the desired requirements for quality, performance and compatibility with other materials and equipment. They are not intended as manufacturing specifications, as this would restrict development, but statements of product performance related to specific methods of testing. British Standards are not legal documents unless referred to in Acts of Parliament, and manufacturers are under no legal obligation to comply with them. However, it is likely that certain British Standards will become the technical criteria for legislation.

British Standard publications

British Standards

These were originally product specifications, but now have wider significance including product quality, testing techniques, acceptable dimensions and recommended use. Manufacturers of articles and equipment claiming compliance with a British Standard may endorse their product with the Institution's certification Kitemark. Specification writing can be considerably reduced by quoting the relevant standard, e.g. BS 12 : 1978 *Portland Cement*, thus eliminating the need for detailed explanations in contract documents. The new codes of practice are now included within this broadened function of British Standards.

Codes of Practice

These are guides to good practice, which may be applied to product design, production, assembly, installation and maintenance. Existing codes have a CP prefix next to their number, e.g. BS CP 114 : Part 2 : 1969. These are gradually being updated for inclusion with British Standards, e.g. BS CP 97 : Part 1: *Code for Metal Scaffolds*, has become BS 5973 : 1981, *Code of Practice for Access and Working Scaffolds*.

Draft for development

These are issued for a maximum of five years when insufficient data about a new product or technique are available. When sufficient information has been accumulated from further research, testing and feedback from industry, the draft becomes a British Standard. Drafts for development are prefixed DD, e.g. DD 51 : 1977, *Guidance on Dimensional Co-ordination in Building*.

Published documents

These are BSI publications which are difficult to locate in the other groups. They are prefixed PD, e.g. PD 6440 : 1969 *Accuracy in Building*.

Acts of Parliament

The following list is a summary of some of the most significant legislation likely to influence design, layout and physical construction of buildings for residential and industrial use.

Ancient Monuments and Archeological Areas Act 1979 and *Field Monuments Act 1972*. Preserve and conserve representative structures, therefore restrict and control development in these areas.

Chronically Sick and Disabled Persons Act 1970. Requires all new public buildings to be designed with regard for disabled people.

Civic Amenities Act 1967. Provides for preservation and conservation of certain buildings and trees, therefore restricts development in some areas.

Control of Office and Industrial Development Act 1965. Prevents overdevelopment in favoured areas such as city centres, and encourages development in areas of high unemployment.

Countryside Acts (Scotland 1967, England and Wales 1968). Provides for access to areas of natural beauty and for limited construction to provide facilities for the public to enjoy these areas.

Defective Premises Act 1972. Emphasises that the builder is responsible for any incorrect or faulty construction in dwellings.

Education Act 1944. Provision of amenities for health and safety in schools and colleges.

Explosives Acts 1875 and 1923. Constructional requirements for structures containing explosive substances.

Factories Act 1961. Extensive document covering the provision of industrial welfare facilities for manual workers.

Fire Precautions Act 1971. Provision of fire communication systems and means of escape in certain buildings not covered by the Factories Act or Offices, Shops and Railway Premises Act.

Food and Drugs Act 1955. Design and construction of buildings where food is processed, stored or sold.

Health and Safety at Work Etc. Act 1974. A relatively new Act which endorses the requirements of many other industrial premises Acts, and provides stricter standards for ensuring the personal health and safety of employees.

Highways Act 1980. Affects the use of roads for site access and contains provision for protection of the public by hoardings and fencing around building sites.

Housing Acts 1974 and 1980. Permits local authority to grant aid for improvement, repair and installation of modern amenities and facilities in substandard housing. Recognises housing action areas for general improvement and reconstruction.

Industry Act 1972. Provision of grants for improving or adapting certain industrial premises in development areas.

Licensing Act 1964. Structural requirements for public houses and other licensed premises.

Mental Health Act 1959. Prohibits registration of premises if accommodation is unsatisfactorily situated or unsympathetically constructed for mental health patients.

New Towns Acts 1959 and 1965. Provides measures for land acquisition, planning and finance of development proposals.

Nursing Homes Act 1963. Requires installation of sufficient facilities and amenities, and high standards of heating and ventilation.

Offices, Shops and Railway Premises Act 1963. Requires adequate standards of health, safety and welfare facilities for employees. Particularly affects temperature, ventilation, lighting, sanitation and fire precautions.

Petroleum Act 1928. Special constructional requirements for buildings storing highly volatile and inflammable liquids.

Prevention of Damage by Pests Act 1949. Structural requirements for prevention of infestation in food storage areas.

Public Health Acts 1939 and 1961. These acts preceded the first Building Regulations of 1965, and they still retain effective control of certain aspects of construction, notably sanitation, drainage and refuse disposal.

Riding Establishments Acts 1964 and 1970. In similarity with other premises accommodating animals for commercial purposes, registration is subject to satisfactory constructional standards.

Slaughter of Animals Act 1958, Slaughter of Poultry Act 1967, Slaughterhouses Act 1958. These cover special design and construction requirements.

Thermal Insulation (Industrial Buildings) Act 1957. Provides local authorities with legislation to reject proposals for inadequately insulated construction. Now largely superseded by Part FF, Building Regulations.

Town and Country Planning Act 1971 and 1972 amendment. Extensive legislation governing all aspects of demolition, development and planning control. Complements the Civic Amenities Act by endorsing preservation of buildings, trees and areas of special significance.

Water Acts, 1945, 1973 and 1981. Provides for creation of area water authority by-laws to originate mandatory procedures for installation of hot- and cold- water services.

Building Regulations

In the UK there are numerous statutory provisions for control of design, construction and use of buildings. The most significant are:

Inner London. London Building Acts 1930–1939 and GLC by-laws provide for independent assessment of technical requirements by the particular area's district surveyor.

England and Wales (except Inner London) 1976. Building Regulations administered by the Department of the Environment and Welsh Office through district councils.

Scotland. Building Standards Regulations 1973 and Building Regulations 1981 administered nationally by the Scottish Development Department and locally by regional councils.

Northern Ireland. Building Regulations made under the Northern Ireland (Temporary Powers) Act 1972. Administered by the Ministry of Finance through district councils.

Index